185 Topics in Current Chemistry

Springer

*Berlin
Heidelberg
New York
Barcelona
Budapest
Hong Kong
London
Milan
Paris
Santa Clara
Singapore
Tokyo*

Electrochemistry VI

Electroorganic Synthesis: Bond Formation at Anode and Cathode

Volume Editor: E. Steckhan

With contributions by
R. D. Little, K. D. Moeller, J.-Y. Nédélec,
J. Périchon, M. K. Schwaebe, I. Tabakovic,
M. Troupel

With 2 Figures and 34 Tables

 Springer

This series presents critical reviews of the present position and future trends in modern chemical research. It is addressed to all research and industrial chemists who wish to keep abreast of advances in the topics covered.

As a rule, contributions are specially commissioned. The editors and publishers will, however, always be pleased to receive suggestions and supplementary information. Reviews are accepted for "Topics in Current Chemistry" in English.

In references Topics in Current Chemistry is abbreviated Top.Curr.Chem. and is cited as a journal.

Springer WWW home page: http://www.springer.de
Visit the TCC home page at http://www.springer.de/

ISSN 0340-1022
ISBN 3-540-61454-0 Springer-Verlag Berlin Heidelberg NewYork

Library of Congress Catalog Card Number 74-644622

This work is subject to copyright. All rights are reserved, whether the whole or part of the material is concerned, specifically the rights of translation, reprinting, reuse of illustrations, recitation, broadcasting, reproduction on microfilm or in any other way, and storage in data banks. Duplication of this publication or parts thereof is permitted only under the provisions of the German Copyright Law of September 9, 1965, in its current version, and permission for use must always be obtained from Springer-Verlag. Violations are liable for prosecution under the German Copyright Law.

© Springer-Verlag Berlin Heidelberg 1997
Printed in Germany

The use of general descriptive names, registered names, trademarks, etc. in this publication does not imply, even in the absence of a specific statement, that such names are exempt from the relevant protective laws and regulations and therefore free for general use.

Typesetting: Macmillan India Ltd., Bangalore-25
SPIN: 10537025   66/3020 – 5 4 3 2 1 0 – Printed on acid-free paper

Volume Editor

Prof. E. Steckhan
Rheinische Friedrich-Wilhelms-Universität Bonn
Institut für Organische Chemie und Biochemie
Gerhard-Domagk-Straße 1
D-53121 Bonn, FRG

Editorial Board

Prof. Dr. *Armin de Meijere*   Institut für Organische Chemie der Georg-August-Universität,
Tammannstraße 2, 37077 Göttingen, FRG
e-mail: ucoc@uni-goettingen.de

Prof. Dr. *K. N. Houk*   University of California, Department
of Chemistry and Biochemistry, 405 Hilgard Avenue,
Los Angeles, CA 90024-1589, USA
e-mail: houk@chem.ucla.edu

Prof. Dr. *Jean-Marie Lehn*   Institut de Chimie, Université de Strasbourg,
1 rue Blaise Pascal, B. P. Z 296/R8, F-67008 Strasbourg Cedex,
France
e-mail: lehn@chimie.u-strasbg.fr

Prof. *Steven V. Ley*   University Chemical Laboratory, Lensfield Road,
CB2 1EW Cambridge, England
e-mail: svl1000@cus.cam.ac.uk

Prof. Dr. *Joachim Thiem*   Institut für Organische Chemie, Universität Hamburg,
Martin-Luther-King-Platz 6, 20146 Hamburg, FRG
e-mail:thiem@chemie.uni-hamburg.de

Prof. *Barry M. Trost*   Department of Chemistry, Stanford University,
Stanford, CA 94305-5080, USA
e-mail: bmtrost@leland.stanford.edu

Prof. Dr. *Fritz Vögtle*   Institut für Organische Chemie und Biochemie der Universität,
Gerhard-Domagk-Straße 1, 53121 Bonn, FRG
e-mail: voegtle@plumbum.chemie.uni-bonn.de

Prof. *Hisashi Yamamoto*   School of Engineering, Nagoya University,
464-01 Chikusa, Nagoya, Japan
e-mail: j45988a@nucc.cc.nagoya-u.ac.jp

Editorial Statement

The series *Topics in Current Chemistry* was founded in 1949 as "Fortschritte der chemischen Forschung" (meaning "advances in chemical research"). The aims and scope at that time read as follows (translation from the German).

"The purpose [of the series] is to report on advances in areas of the chemical sciences of high current interest. The principal goal is not the exhaustive description of all published results, but rather the critical examination of the literature and the elucidation of the main directions that developments are taking. Likewise, these progress reports are not written exclusively for specialists, but rather for all interested chemists, who want to inform themselves about areas bordering on their work."

This concept and the publication of topical volumes from the beginning showed an amazing foresight on the part of the founding editors. Over the years progress in the chemical sciences has been enormous, and the volume of primary publications per year has increased tremendously. At the same time the application of the powerful tools of chemical investigation to other fields has led to a growing understanding of the chemical basis of the natural, medical and material sciences, with the result that the chemical sciences have become increasingly interdisciplinary. Through all these changes the original concept and organization of *Topics in Current Chemistry* has allowed it to continue to play an important role in the secondary literature. Indeed, with the rapidly changing landscape of the chemical sciences, the need for such a series is ever increasing.

As we look to the future, the challenges and opportunities facing the chemical research community are great. It is the goal of the current editorial board to continue to provide that community with topical volumes containing critical reviews of the present and future trends across the entire range of the modern chemical sciences. We join Springer-Verlag in thanking the editors now leaving the editorial board, Jack D. Dunitz, Klaus Hafner, Sho Ito, Kenneth N. Raymond and Charles W. Rees, whose efforts over the last decades served to maintain the high standards of this series.

Armin de Meijere, Göttingen
Kendall N. Houk, Los Angeles
Jean-Marie Lehn, Strasbourg
Steven V. Ley, Cambridge, UK

Joachim Thiem, Hamburg
Barry M. Trost, Stanford
Fritz Vögtle, Bonn
Hisashi Yamamoto, Nagoya

Preface

In organic chemistry the strategic steps in complex syntheses are bond forming reactions. Therefore, as mature method, electroorganic has to compete with the powerful repertoire of alternative synthetic processes which have been developed for this pupose in recent years. The volume Electrochemistry VI of Topics in Current Chemistry reviews the state of the art of modern bond forming reactions at the anode and cathode, demonstrating that they continue to be highly competitive with alternative methods. It highlights both the current value and the large potential of organic electrochemistry for the selective formation of carbon-carbon and carbon-heteroatom bonds and for the generation of complex organic molecules using electrochemical key steps. The utility of electrochemistry for the generation of highly reactive intermediates under mild conditions and for the selective reversal of the polarity of functional groups is demonstrated for ring forming reactions and carbon-carbon cross-coupling reactions. The contents range from the synthesis of natural products to the preparation of pharmaceuticals, from the generation of unsymmetrical biaryls to the construction of peptide mimetics. The electrochemical prodedures are critically compared with alternative chemical processes and mechanistic considerations are included. As guest editor I am very thankful that again well known experts in their fields agreed to contribute four concise reviews to this volume.

Bonn, October 1996   Prof. Steckhan

Table of Contents

Reductive Cyclizations at the Cathode
R. D. Little and M. K. Schwaebe   1

Intramolecular Carbon – Carbon Bond Forming Reactions at the Anode
K. D. Moeller 49

Anodic Synthesis of Heterocyclic Compounds
I. Tabakovic 87

Organic Electroreductive Coupling Reactions Using Transition Metal Complexes as Catalysts
J.-Y. Nédélec, J. Périchon and M. Troupel 141

Author Index Volumes 151–185 175

Table of Contents of Volume 170

Electrochemistry V

Electrochemical Reactions of Fluoro Organic Compounds
T. Fuchigami

Electrochemical Reactions of Organosilicon Compounds
J. Yoshida

Electroenzymatic Synthesis
E. Steckhan

Electrochemistry for a Better Environment
P. M. Bersier, L. Carlsson and J. Bersier

Reductive Cyclizations at the Cathode

R. Daniel Little and Michael K. Schwaebe

Department of Chemistry, University of California, Santa Barbara, CA 93106, USA

Table of Contents

1 Introduction	2
2 Electrohydrocyclization	3
2.1 An Application to Natural Products Synthesis: Sterpurene ...	4
2.2 Allene Tethered to α,β-Unsaturated Ester	6
2.3 Variations on a Theme: Bis-Imine Cyclization	7
3 Variations on a Theme	8
3.1 Electroreductive Cyclization: Electron Deficient Alkene/Carbonyl	8
3.1.1 Mechanism	9
3.1.2 An Approach to Pentalenolactone E Methyl Ester	11
3.1.3 Formal Total Synthesis of Quadrone	13
3.1.4 The [3.2.1] Framework: Scope and Limitations	15
3.2 Non-Conjugated Alkene/Carbonyl	18
3.3 Alkene/Ester	21
3.4 Allene and Alkyne Acceptors/Carbonyl	21
3.5 Nitrile/Carbonyl	22
3.6 The Aromatic/Carbonyl Variation	26
3.7 Pyridinium/Carbonyl	29
4 Tandem Cyclizations	31
4.1 Tandem Cyclization of Non-Conjugated Dienones	31
4.2 Enol Phosphates of 1,3-Dicarbonyl Compounds	32
4.3 Tandem Cyclization of Enol Phosphates	34

5 Displacements	35
6 Halides	37
6.1 Cyclizations Involving Alkylidene Malonates	37
6.2 Vitamin B_{12}-Mediated Cyclizations	38
6.3 Alkynyl Halides	40
6.4 Allylic and Aryl Halides	41
6.5 Cyclization of 1,n-Dihalides	43
7 Ring Expansion	44
8 Reduction of Sulfonamides: Cyclization Promoted by N-Tosyl Group Cleavage	45
9 Concluding Remarks	46
10 References	46

In a recent publication, Schäfer and coworkers point out the utility of the electrode as a "reagent" which is effective in promoting bond formation between functional groups of the same reactivity, or polarity. They accurately note that reduction at a cathode, or oxidation at an anode, renders electron-poor sites rich, and electron-rich sites, poor. This reactivity-profile reversal clearly provides many opportunities for the formation of new bonds between sites formally possessing the same polarity, *provided* only one of the two groups is reduced or oxidized. Electrochemistry provides an ideal solution to the issue of selectivity, given that a controlled potential reduction or oxidation is readily achieved using an inexpensive potentiostat. In this article we address cyclizations that are initiated by reduction at a cathode. A wide variety of ring systems have been constructed in this manner.

1 Introduction

In a recent publication, Schäfer and coworkers point out the utility of the electrode as a "reagent" which is effective in promoting bond formation between functional groups of the same reactivity or polarity [1]. They accurately note that reduction at a cathode, or oxidation at an anode, renders electron-poor sites rich, and electron-rich sites poor. For example, reduction of an α,β-unsaturated ketone leads to a radical anion where the β-carbon possesses nucleophilic rather than electrophilic character. Similarly, oxidation of an enol ether affords a radical cation wherein the β-carbon displays electrophilic, rather than its usual nucleophilic behavior [2]. This reactivity-profile reversal clearly provides many opportunities for the formation of new bonds between sites formally possessing the same polarity, *provided* only one of the two groups is reduced or oxidized. Electrochemistry provides an ideal solution to the issue of selectivity, given that a controlled potential reduction or oxidation is readily achieved using an inexpensive potentiostat.

Electrochemical processes are conducted under what is referred to as either "constant current" (CC) or "controlled potential" (CP) conditions [3–7]. The former is often preferred because it is less expensive to implement, since it does not require the acquisition of a potentiometer, is amenable to scaleup, and the reaction times are often quite short. The disadvantage is that, in order to maintain a constant current, the potential must change, becoming more negative or positive, depending upon whether a reduction or an oxidation is being investigated. Controlled potential methods allow one to set the potential to a value corresponding to that of the electrophore, much the same way that a light filter allows one to irradiate a particular chromophore selectively in a photochemical process. And, just as one obtains a UV spectrum to determine the appropriate filter prior to conducting photolysis, so one obtains the analogous cyclic voltammogram (CV) corresponding to the material to be studied electrochemically. It describes how the current changes as a function of potential.

In this article we address cyclizations that are initiated by reduction at a cathode. A wide variety of ring systems have been constructed in this manner. Given its importance in the development of organic electrochemistry, it is appropriate to begin our discussion with the electrohydrocyclization (EHC) reaction pioneered in the 1960s by Baizer and co-workers [8–10].

2 Electrohydrocyclization

Without doubt, this is the best known cathodically-initiated cyclization. The transformation corresponds to the intramolecular variation of electrohydrodimerization [11–13], a process well known for its application to the preparation of adiponitrile [14].

$$(R_2C)_n \begin{matrix} \overset{\beta}{CH=CHX} \\ CH=CHX \\ \underset{\beta'}{} \end{matrix} \quad \xrightarrow[\text{Et}_4\text{NOTs, CH}_3\text{CN/H}_2\text{O}]{-1.8 \text{ to } -2.1 \text{ V (Hg; SCE)}} \quad (R_2C)_n \begin{matrix} CH_2X \\ CH_2X \end{matrix} \qquad (1)$$

    1   **2**

Substrates for EHC reactions consist of two electron-deficient alkenes tethered to one another. Reduction leads to the formation of an adduct wherein the β-carbons are joined by a new sigma bond. As illustrated in Table 1, the methodology is exceptionally useful for the construction of three-, five-, and six-membered rings, but not for rings of sizes seven and eight.

When cyclization is slow, either because the size of the ring to be formed is unfavorable (e.g., n = 6 and 8), or the β-carbon of the substrate is sterically encumbered, the yield of cyclized adduct drops precipitously, and saturation of the C=C π bond becomes an important side reaction. Heterocyclic systems can

Table 1. Electrohydrocyclization

n	Ring size	R	X	Yield (%)
1	3	Et	CO_2Et	98
2	4	H	CO_2Et	41
2	4	H	CN	15
3	5	H	CO_2Et	~ quant
4	6	H	CO_2Et	90
5	7	H	CO_2Et	~ 10
6	8	H	CO_2Et	0

also be obtained in high yield, as the following examples illustrate [8, 10, 15].

$$\mathbf{3} \xrightarrow[\text{Et}_4\text{NOTs (86\%)}]{\text{2e, CH}_3\text{CN/H}_2\text{O}} \mathbf{4} \tag{2}$$

$$\mathbf{5} \xrightarrow[\text{Et}_4\text{NOTs (89\%)}]{\text{2e, CH}_3\text{CN/H}_2\text{O}} \mathbf{6} \tag{3}$$

A generally accepted mechanism for EHC reactions involves reduction to a radical anion, followed by protonation, and cyclization of the resulting radical; the addition of a second electron and proton completes the event [9, 16]. It has also been suggested that electron transfer and cyclization are concerted processes. This is based on the fact that the polarographic half-wave potential for the systems shown in Table 1 are shifted to more positive values for systems that undergo cyclization than for those that do not. In the latter instance, the half-wave potential is essentially that of an isolated electron-deficient alkene. Furthermore, cyclic voltammograms do not display an oxidation wave on the reverse scan. Clearly, on the timescale of the measurement (polarographic or CV), a follow-up reaction occurs more rapidly than reverse electron transfer [4]. The fact that the potential shift is not observed in systems that fail to close adds credibility to the idea that the rapid follow-up process might be cyclization, as originally suggested.

2.1 An Application to Natural Products Synthesis: Sterpurene

A relatively recent application to natural product synthesis stems from efforts to synthesize a sesquiterpene called 1-sterpurene **7** [17]. This substance is thought to be the causative agent of the so-called "silver leaf disease" that affects certain species of shrubs and trees. The strategy focuses on three key steps: (a) electrochemical cyclization of the bis unsaturated ester **11** to produce the five-membered ring of **10**, (b) a Ruhlman-modified acyloin condensation to

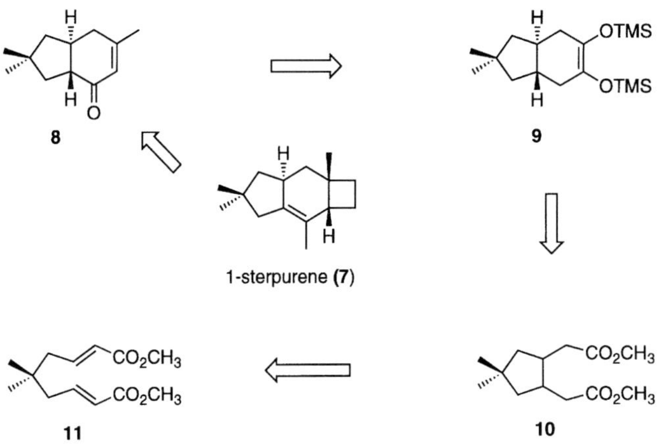

Scheme 1

provide the six-membered ring (**10** to **9**), and (c) a photo[2 + 2] cycloaddition to generate the third.

The electrohydrocyclization of **11** was investigated under a number of different conditions. As noted in Table 2, both the yield and stereoselectivity varied in response to changes in electrode and proton donor. Environmental factors clearly make the use of a glassy carbon electrode preferable. Yields range from 58 to 87%, and stereoselectivity from a low of 2.6:1 to as high as 14.8:1, the latter occurring when the reaction was conducted in the presence of cerium (III) chloride (see last entry of Table 2). In each instance the *trans* isomer **10a** was preferred. The influence of cerium chloride, while interesting and significant in terms of its effect, is not clearly understood. The intent was that it should complex with the substrate to form a template which might favor the formation of one stereoisomer. While this may be so given the results, there is not sufficient data to support the claim with confidence. Additional study is needed.

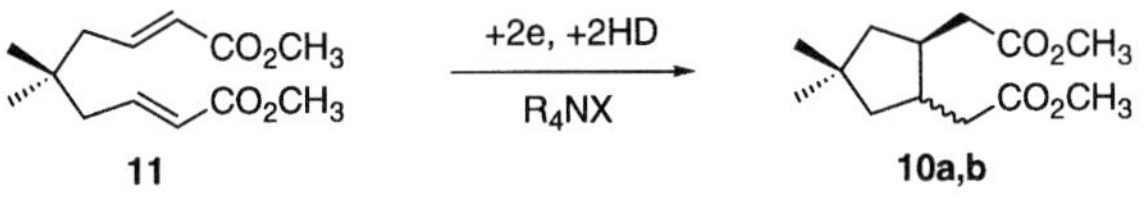

$$\mathbf{11} \xrightarrow[R_4NX]{+2e,\ +2HD} \mathbf{10a,b} \quad (4)$$

Table 2. Effect of proton donor and electrode material

Electrode	Proton source	Trans/cis ratio	Yield (%)	Additive[a]
Hg	AcOH, H$_2$O	2.6:1	82–87	none
Hg	CH$_2$(CO$_2$Et)$_2$	7.5:1	66	none
glassy carbon	CH$_2$(CO$_2$Et)$_2$	7.1:1	73	none
Cu	CH$_2$(CO$_2$Et)$_2$	11.6:1	58	none
Hg	CH$_2$(CO$_2$Et)$_2$	14.8:1	73	1.3 equiv CeCl$_3$

[a] Suspension in acetonitrile

This example aside, the utility of the electrohydrocyclization reaction in the assembly of natural products remains essentially untapped. The reason for this observation is not entirely clear. Consider, for example, the rapid and efficient construction of the perhydrophenanthrene skeleton as exemplified by the conversion illustrated below [18]. The reaction is stereospecific, leading to the *trans-anti-trans* ring fused adduct **13** in a 65 to 72% yield. It takes little thought to imagine the application of this powerful transformation to the synthesis of steroids. Yet, this opportunity has not been realized.

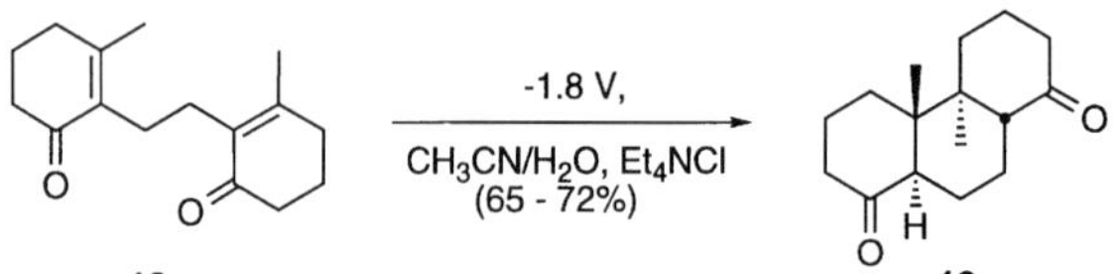

$$\mathbf{12} \xrightarrow[\substack{CH_3CN/H_2O,\ Et_4NCl \\ (65 - 72\%)}]{-1.8\ V,} \mathbf{13} \tag{5}$$

2.2 Allene Tethered to α,β-Unsaturated Ester

An exceptionally interesting example of the electrohydrocyclization reaction involves the use of allenes which are tethered to α,β-unsaturated esters (Table 3) [19]. The chemistry takes place in a manner wherein the new carbon–carbon bond forms between the central carbon of the allene and the β-carbon of the unsaturated ester. Of particular value is the preservation of one of the double bonds of the original allene, thereby providing functionality for further elaboration. It is important to carry out these transformations in an undivided cell, as the use of a divided cell led to hydrogenation of the olefins instead of cyclization.

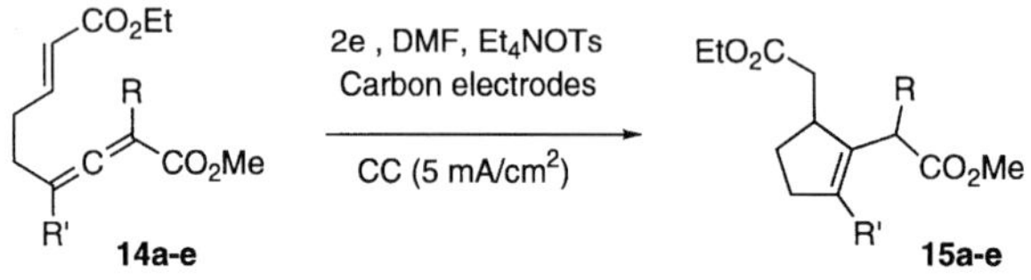

$$\mathbf{14a\text{-}e} \xrightarrow[\substack{2e,\ DMF,\ Et_4NOTs \\ \text{Carbon electrodes} \\ CC\ (5\ mA/cm^2)}]{} \mathbf{15a\text{-}e} \tag{6}$$

Table 3. EHC of allenes tethered to unsaturated esters

	R	R'	F/mol	Yield (%)
a,	n-Bu-	H	8.4	96
b,	$(CH_3)_2C(OH)$-	H	8.7	64
c,	$THPOCH_2$	Me	11.3	69
d,	H	Me	4.5	93
e,	$CH_3C(OCH_2CH_2O)CH_2CMe_2CH_2$-	Me	5.8	95

2.3 Variations on a Theme: Bis-Imine Cyclization

It is easy to imagine variations on the electrohydrocyclization theme. Consider, for example, the possibility of replacing both of the electron-deficient alkenes with imines. The chemistry of these systems has been explored and provides a convenient route to piperazines [20]; note Table 4 and equation 7. Yields range from poor (38%) to exceptionally high (95%). Lead, zinc, tin, or platinum electrodes work satisfactorily, though the yields tend to be slightly higher using lead. The transformations are carried out in DMF as the solvent, using methanesulfonic acid as a proton donor.

$$16\text{a-e} \xrightarrow[\text{MsOH/DMF}]{2e\ (\text{Pb})} [17] \longrightarrow 18\text{a-e} \quad (7)$$

Table 4. Electrohydrocyclization of *bis*-imines

	Ar	R	R'	Yield (%)
a,	C_6H_5	H	H	95
b,	$p\text{-MeOC}_6H_4$	H	H	82
c,	$o\text{-HOC}_6H_4$	H	H	42
d,	$o\text{-HOC}_6H_4$	i-BuS	H	38
e,	$o\text{-HOC}_6H_4$	$-(CH_2)_4-$		78

In the absence of the acid, cyclization does not occur. N,N'-Dibenzylidenethylenediamine (**16a**), for example, is converted to N,N'-dibenzylethylenediamine (**19**) in $> 60\%$ yield in its absence.

$$\mathbf{16a} \xrightarrow[(>60\%)]{2e,\ \text{DMF}} \mathbf{19} \quad (8)$$

While the yields are generally good for the cyclization of diimines derived from aldehydes, the process does not work satisfactorily using diketoimines such as **20**. This is not too surprising, given the need to form two contiguous quaternary centers in **21**.

$$\mathbf{20} \xrightarrow[\substack{\text{MsOH/DMF}\\(20\%)}]{2e\ (\text{Pb})} \mathbf{21} \quad (9)$$

The method lends itself nicely to the construction of enantiomerically enriched tri- and tetrasubstituted piperazines such as **23**, substances that prove useful as ligands in enantioselective synthesis, such as that illustrated in Eq. (11) [20].

$$\text{structure 22} \xrightarrow[\substack{\text{3. BnBr} \\ \text{4. NaH, BnBr}}]{\substack{\text{1. 2e, MsOH/DMF} \\ \text{2. TMSCl, Et}_3\text{N}}} \text{structure 23} \qquad (10)$$

$$\text{pentyl-CHO} + \text{Et}_2\text{Zn} \xrightarrow[\text{(91\%, 81\% ee)}]{\text{23 (cat), PhCH}_3} \text{24} \qquad (11)$$

As noted previously, many of the cathodic cyclizations discussed in this article are variations on the electrohydrocyclization theme developed by Baizer and coworkers [8–14, 16, 17, 21]. The next section of this article, for example, deals with what has been referred to as the electroreductive cyclization (ERC) reaction, a process that leads to cyclization between an electron-deficient alkene and an aldehyde or ketone. With this thought in mind, several of the section titles are formulated to highlight the functional groups to be joined; the following is representative.

3 Variations on a Theme

3.1 Electroreductive Cyclization: Electron Deficient Alkene/Carbonyl

As indicated, these transformations lead to the formation of a new sigma bond between two formally electron-deficient centers [4, 22, 23], in this instance between the β-carbon of an electron-deficient alkene and a carbonyl carbon.

$$\textbf{25} \xrightarrow[\substack{\text{(when G = COR')} \\ \text{EWG = CO}_2\text{R, COR, CN, etc.} \\ \text{G = COR', CH=CREWG} \\ \text{HD = proton donor (see text)}}]{\text{2e, 2HD}} \textbf{26} \qquad (12)$$

Overall, the process requires the consumption of two electrons and two protons. The structure and acidity of effective proton donors vary from mineral to carbon acids; often, a simple dialkyl malonate is effective. It is easy to monitor current consumption using a simple, commercially available coulometer [3, 4].

When cyclization occurs between an electron-deficient allene and a ketone, as is the case with **27**, butenolides are produced [19]. Given the importance of this functional group in cardiac glycoside natural products, the simplicity of the starting materials, and the facility of the cathodic cyclization, it may be worth noting the opportunity to utilize this transformation in the construction of these natural products.

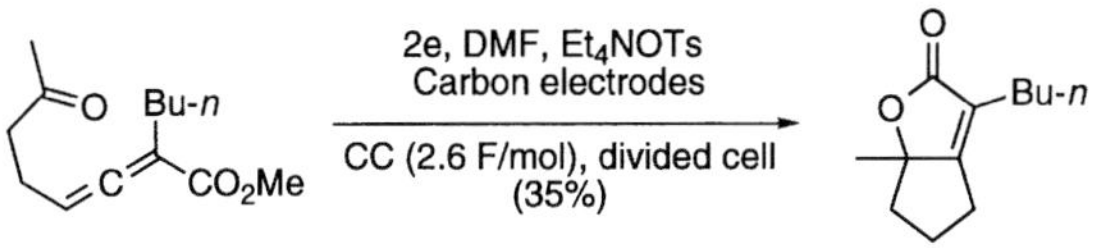

$$\text{27} \xrightarrow[\text{CC (2.6 F/mol), divided cell}]{\text{2e, DMF, Et}_4\text{NOTs, Carbon electrodes}} \text{28} \quad (35\%) \tag{13}$$

3.1.1 Mechanism

The mechanism of the electroreductive cyclization reaction has been studied in some detail [22]. The initial thought was that it occurred via the cyclization of the radical anion derived, for example, from **25** in the first reduction step. A moment's reflection, however, reveals that there are many more mechanistically viable pathways, especially when one realizes that the transformation involves five steps – two electron transfers (symbolized below by "e" and "d", the latter corresponding to a homogeneous process), two protonations ("p"), and cyclization ("c"). In principle, these could occur in any order, and with any one of the steps being rate-determining.

Of hundreds of theoretically possible pathways, the list can be trimmed to four using linear sweep voltammetry (LSV) and chemical arguments [22]. The LSV method is an exceptionally powerful one for analyzing electrochemical processes [24–27]. From LSV studies, it was concluded that a single heterogeneous electron transfer precedes the rate-determining step, cyclization is first order in substrate, and that proton transfer occurs before or in the rate-determining step. The candidates include: (a) e-c-P-d-p (radical anion closure),

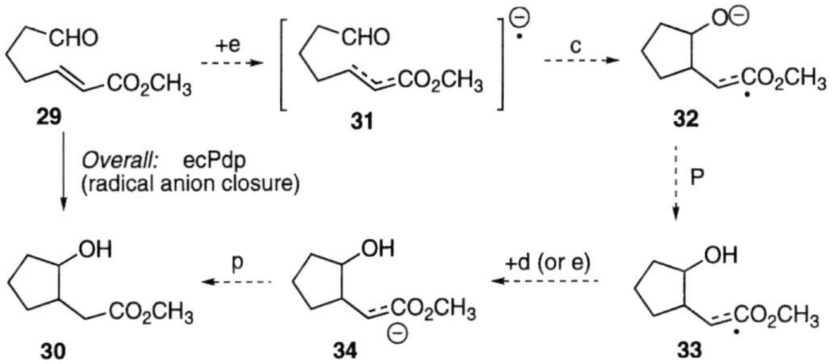

Scheme 2

(b) e-P-c-d-p (radical closure), (c) e-p-C-d-p (radical closure), and (d) e-P-d-c-p (anion closure). The first of these cases is portrayed in Scheme 2 to illustrate the use of the notation; when a symbol appears as a capital letter, it refers to the rate-determining step.

To decide between these possibilities, compound **35** was designed and synthesized; in principle, this system allows an equal opportunity for closure onto the alkene and/or the aldehyde [22].

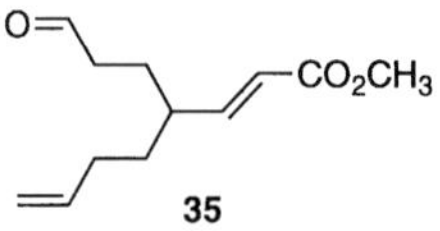

Consider first, the possibility of radical cyclization involving **36**, it being formed after electron transfer ("e") and rate determining protonation ("P"). Clearly, **36** has the option of undergoing either a 5-*exo-trig* cyclization onto an alkene or an aldehyde, with both rate constants approximating $10^5/s$ [28–31]. The literature indicates that the former process is irreversible, while ring opening of the oxygen-centered radical **37** occurs with a rate constant of $\sim 10^8/s$ [28–31]. Should **38** form, it follows that at least some of aldehyde **39** ought to be detected.

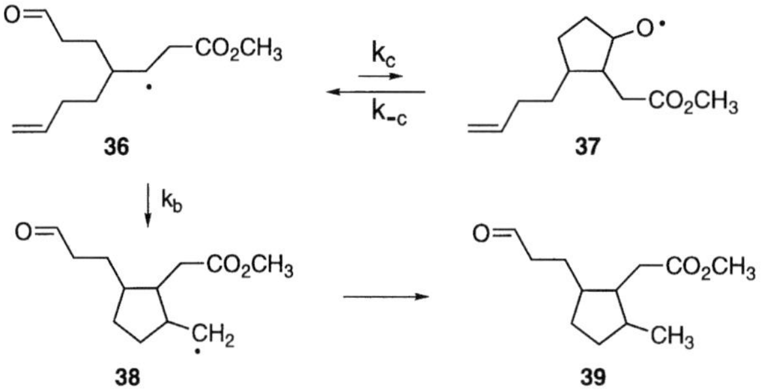

Scheme 3

In practice, reduction of **35** (-2.43 V vs SCE) in the presence of 3,5-dimethylphenol as a proton donor, tetra-*n*-butylammonium hexafluorophosphate as the supporting electrolyte, and DMF as the solvent, led to the γ-hydroxy ester **40** and lactone **41** [22]. No sign of any material resulting from cyclization onto the alkene was detected. It was concluded that radical cyclization does not occur in this instance, and that the homogeneous electron transfer rate exceeds that of a 5-*exo-trig* radical cyclization, thereby implying the operation of either a radical anion or carbanion cyclization pathway.

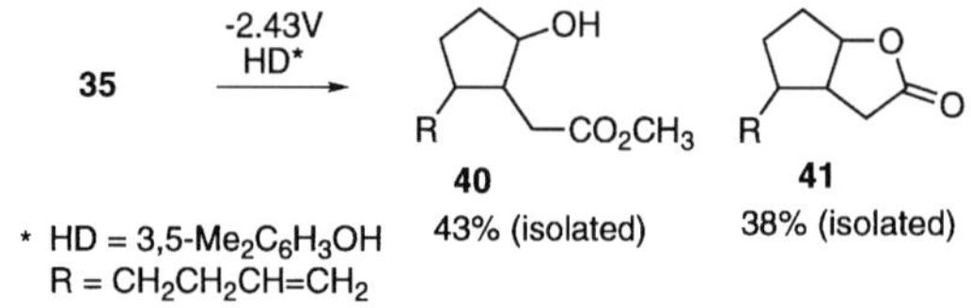

(14)

* HD = 3,5-Me$_2$C$_6$H$_3$OH
R = CH$_2$CH$_2$CH=CH$_2$

40 43% (isolated)
41 38% (isolated)

The radical anion pathway (e-c-P-d-p Scheme 2) requires a rate-determining protonation *after* cyclization, i.e., a slow protonation of a hard oxyanion. However, such proton transfer rates are usually diffusion controlled, so this seems unlikely [32, 33]. On the other hand, the carbanion closure (e-P-d-c-p) portrayed in Scheme 4 requires a very reasonable suggestion that the rate-determining step corresponds to protonation of the soft, weakly basic radical anion **42**, prior to cyclization [32–35]; this is the preferred mechanism. One must use caution, however, realizing that these conclusions are drawn for the particular set of substrates which were examined. In some cases, radical anion cyclization remains a viable option.

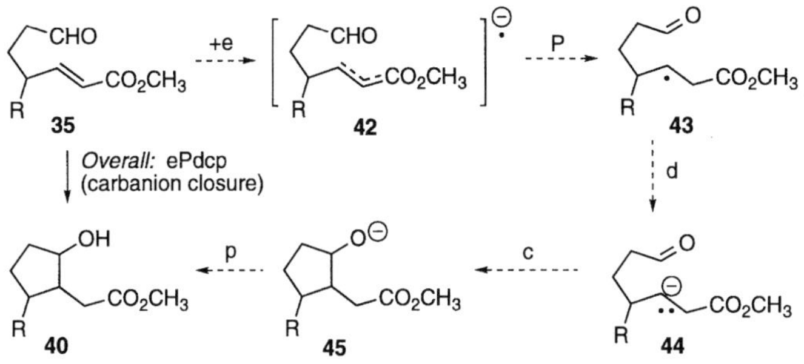

Scheme 4

Unlike electrohydrocyclization, the ERC reaction has seen more widespread use as an important step in the assembly of naturally occurring materials. Several of the efforts to do so are outlined in the following sections.

3.1.2 An Approach to Pentalenolactone E Methyl Ester

Pentalenolactone E methyl ester (**46**), an angularly fused sesquiterpene lactone, was first isolated and characterized by Cane and Rossi [38]. One approach to the synthesis of this material is illustrated in Scheme 5. Key to the successful implementation of the plan is the synthesis of butenolide **49**, the electrochemically promoted cyclization of **49** to the tricyclic γ-lactone **48**, ring opening of the latter to convert the linearly fused system to the angularly fused six-membered ring lactone **47**, and functional group elaboration leading to the natural product **46** [36, 37].

Both butenolides **50** and **51**, one tethered to an α,β-unsaturated ester, the other to an unsaturated nitrile, failed to undergo cyclization. Instead, the C=C π-bond of the butenolide was reduced, leading cleanly and efficiently to saturation of that bond, and to compounds **54** and **55**, respectively. In contrast, the corresponding alkylidene malonate **52**, as well as the alkylidene malononitrile **53**, both cyclize to afford a mixture of the *cis-anti-cis* and *cis-syn-cis* linearly fused lactones **56** and **57**.

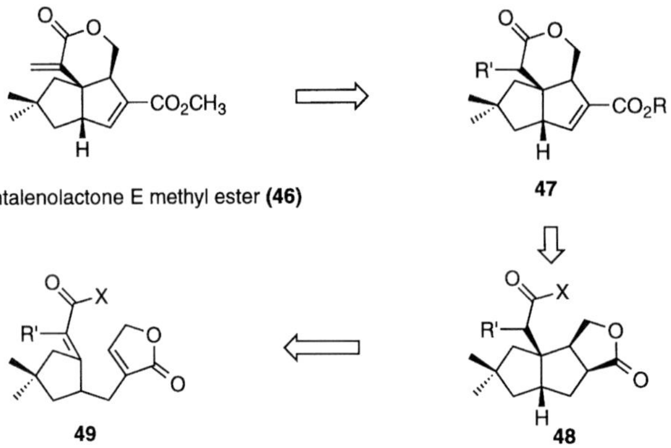

pentalenolactone E methyl ester **(46)** **47** **48** **49**

Scheme 5

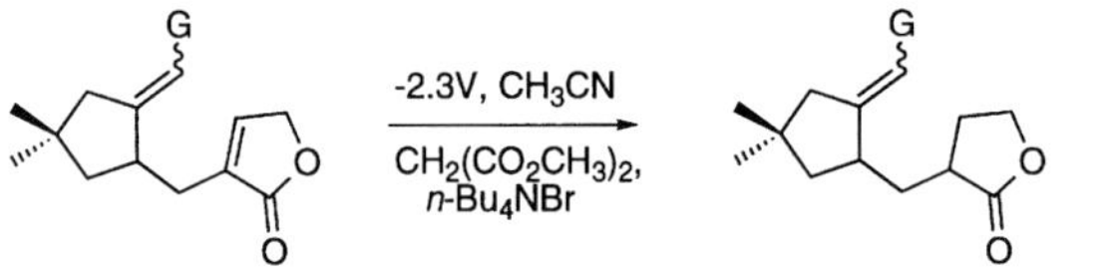

50, G = CO_2CH_3
51, G = CN

54, G = CO_2CH_3, 84%
55, G = CN, >95%

(15)

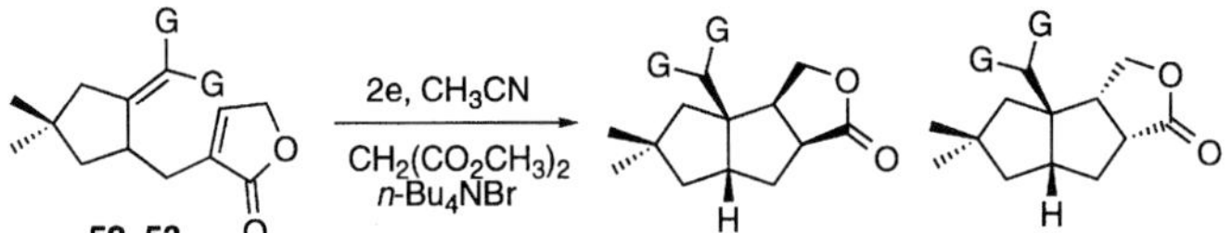

52, 53 **56** **57**

(16)

	G	potential (V vs SCE)	combined yield (%)	ratio (**56/57**)
52,	CO_2CH_3	-2.1	90	1:1
53,	CN	-1.6	23	1:1

That **52** and **53** undergo cyclization, while **50** and **51** do not, points to one of the significant advantages of controlled potential electrochemical methodology over alternative, non-electrochemical methods [37]. That is, that one can vary the potential to match the electrophore. In this instance, when the butenolide is the easier electrophore to reduce, cyclization fails. By switching from a mono to a doubly activated electrophore, the roles reverse. Cyclization occurs from the latter onto the butenolide. Presumably, the reason for this behavior is related to the higher basicity of the radical anion which is produced upon reduction of **50** and **51** compared to that of the more highly delocalized and presumably less basic species resulting from the reduction of **52** and **53**. In the former cases, acid-base chemistry dominates cyclization.

Another interesting feature of the electrochemistry of the doubly activated systems **52** and **53** can be discerned from the data presented in Table 5. First, the dinitrile is easier to reduce than the malonate. While this is a common trait of such systems, the difference in peak potentials is striking, corresponding to as much as 0.5 V, or ~11 kcal. Second, the stereoselectivity changes significantly depending upon the choice of supporting electrolyte, varying from zero (i.e., a 1/1 mixture of stereoisomers) when n-Bu_4NBr is used, to as high as 11/1 (**56/57**) in the presence of $Mg(ClO_4)_2$. Finally, we note that the major product corresponds to the *cis-anti-cis* adduct **56**, as required in the construction of pentalenolactone E methyl ester (**46**).

Table 5. Influence of supporting electrolyte on stereoselectivity in the reductive cyclization of **52** and **53**

G	Potential (V vs SCE)	Combined yield (%)	Ratio	Electrolyte
CO_2CH_3	−2.1	90	1/1	n-Bu_4NBr
CN	−1.6	23 (+25% starting material)	1/1	n-Bu_4NBr
CN	−1.7	77	3/1	$LiClO_4$
CN	−1.6	62	11/1	$Mg(ClO_4)_2$

The stereoselectivity may have its origin in the ability of lithium and magnesium to serve as coordinating metals. For example, in the case of **53** undergoing cyclization in the presence of magnesium perchlorate, it seems reasonable to postulate the existence of an intermediate such as **58**, where the metal is associated with the butenolide as well as the reduced alkylidene malononitrile. Metal coordination to the nitrogen end of the CN-unit in the intermediate is precedented in carbanion chemistry [39]. This effectively places the beta carbon above that of the corresponding carbon in the butenolide. Sigma bond formation therefore establishes all but one stereocenter, it being determined in the final protonation of the enolate. That it should occur to afford a *cis* ring fusion is entirely reasonable given the substantial energy difference between *cis* and *trans* fused bicyclo[3.3.0] ring systems [36, 40].

(17)

3.1.3 Formal Total Synthesis of Quadrone

The utility of electroreductive cyclization chemistry is demonstrated quite nicely in its application to a formal total synthesis of quadrone (**59**) [41]. This fungal metabolite isolated from *Aspergillus terreus*, displays in vivo and in vitro cytotoxicity. One approach focuses upon three transformations, two involving sigma bond formation via electroreductive cyclization (ERC) and serving to convert **60** to **61** and **62** to **63**, the third being an oxidative cyclization used to generate **64**.

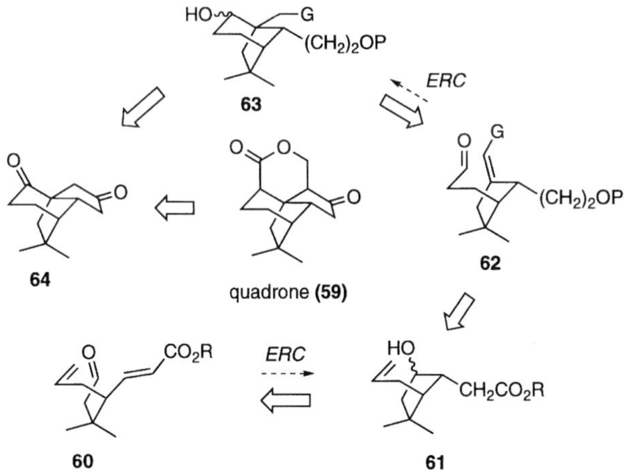

Scheme 6

Controlled potential reduction of **60** in the presence of dimethyl malonate as a proton donor afforded a mixture of two products, the γ-hydroxy ester **65** and lactone **66**, in a combined yield of 89%; each was converted to a common intermediate, **67**.

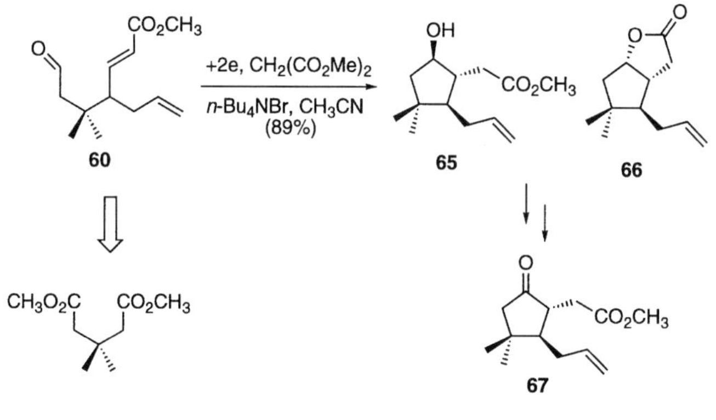

(18)

One can attribute the selective formation of materials with the ester and allyl units *trans* to one another, to the preference for the allyl unit to occupy a pseudoequatorial rather than a pseudoaxial orientation in the product-determining transition state. Compare, for example, transition state formulation **68** with **69**. This stereochemical outcome is fortunate, as later on in the sequence, it is necessary for the allyl unit (after functional group modification) to swing across the top face of the cyclopentyl ring system during the conversion of **62** to **63**. Were the substituents *cis* to one another, this would not be possible.

[structure 68] vs [structure 69]

* = anion or radical anion

Keto ester **67** was converted to the unsaturated nitrile **70** in a routine manner. The latter proved to be an exceptionally useful intermediate. Concern that the significant steric demands which are associated with the formation of a sigma bond to the fully substituted beta carbon of the unsaturated nitrile would prevent reaction from occurring, were allayed by the discovery that the controlled potential reduction of **70** at -2.4 V in the presence of dimethyl malonate as the proton donor, afforded a 90% isolated yield of the requisite [3.2.1] adduct **71**. This material was subsequently converted to enone **72** [42], a convergent point with an existing synthesis of quadrone (**59**).

$$\mathbf{70} \xrightarrow[\substack{n\text{-Bu}_4\text{NBr, CH}_3\text{CN} \\ (90\%)}]{-2.4\text{V, CH}_2(\text{CO}_2\text{Me})_2} \mathbf{71}, P = \text{SiPh}_2\text{Bu-}t \quad (19)$$

3.1.4 The [3.2.1] Framework: Scope and Limitations

As indicated by the conversion of **70** to **71**, the electroreductive cyclization reaction provides as excellent method for the assembly of the bicyclo[3.2.1]octane ring system. Several additional examples are portrayed in the following equations. In general, the use of an unsaturated nitrile rather than the corresponding ester is preferred, as this precludes lactone formation, and therefore reduces the number of possible products.

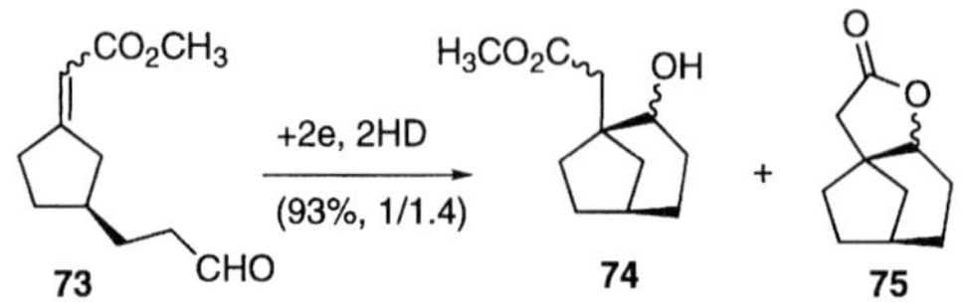

$$\text{73} \xrightarrow{+2e, 2HD} \text{74} + \text{75} \quad (20)$$
(93%, 1/1.4)

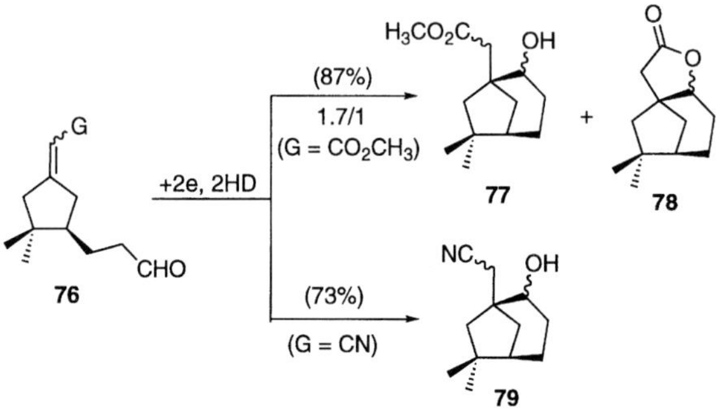

$$\text{76} \xrightarrow{+2e, 2HD} \begin{array}{c} (87\%) \\ 1.7/1 \\ (G = CO_2CH_3) \end{array} \text{77} + \text{78} \quad (21)$$

$$\xrightarrow{(73\%)}_{(G = CN)} \text{79} \quad (22)$$

As with any transformation, this methodology is not without limitations. For example, the γ-hydroxy unsaturated ester **80** failed to undergo the desired cyclization [43]. Instead, elimination occurred, presumably leading to the extended enolate **82** which condenses onto the aldehyde unit to afford the [3.3.0] adduct **83** in a 45% yield.

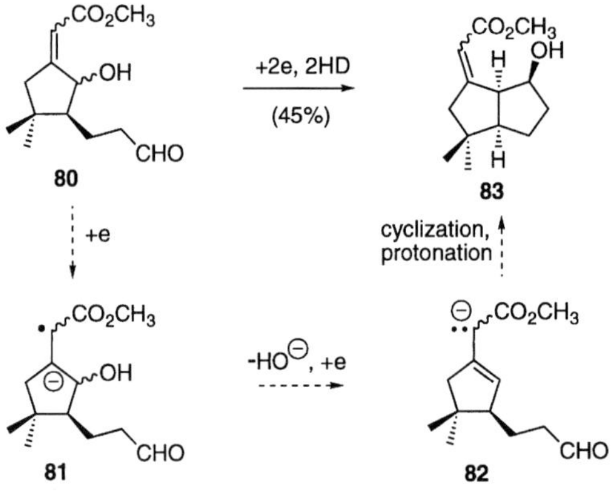

Scheme 7

Another occasionally troublesome process is related to the tendency of some substrates to preferentially, or competitively, undergo acid-base chemistry [43,44]. Of course, this is not surprising, considering the nature of the putative intermediates. One system where this process diminishes the efficiency of the electroreductive cyclization is that of compound **84**. In this reaction, no more than 35–40% of cyclized material could be isolated, in this case lactone **86**. Based on the experiments described below, it is presumed that the saturated ester serves as a proton source which is able to quench the intermediate **85** responsible for cyclization. When this proton source is rendered unavailable by conversion of the side chain ester to an ether as with compounds **87** and **89**, the yield for cyclization increases to 54 and 72%, respectively.

Another example of a case where acid-base chemistry competes with cyclization is found in efforts to construct an analog of the Corey lactone [45]. The enantiomerically pure unsaturated ester **91** was assembled and subjected to the conditions indicated in Eq. (26). In this instance, dimethyl methylmalonate was used as the proton donor to avoid 1,4-addition of the conjugate base to **91**. Cyclization afforded a combined 77% isolated yield of the γ-hydroxy ester **92** and the lactone **93**; the former could be converted to the lactone in the straightforward manner portrayed above. Unfortunately, these materials were accompanied by the formation of the conjugated diene **94**, a substance which undoubtedly arises via the intervention of acid-base chemistry, leading to β-elimination of the silyloxy unit.

3.2 Non-Conjugated Alkene/Carbonyl

A powerful method to convert non-conjugated keto olefins into cyclic alcohols [46, 47] utilizes the cathodic reduction of a ketone to afford a ketyl, which subsequently undergoes cyclization onto the pendant alkene. As illustrated in Eqs. (27)–(30), the process provides access to mono-, bi-, and heterocyclic systems from very simple starting materials. It is especially well-suited to the construction of five-membered rings, less so for six-, and is not effective in producing seven-membered rings.

$$\mathbf{95} \xrightarrow[\text{MeOH/dioxane} \atop (98\%)]{e\ (C\ rod),\ Et_4NOTs} \mathbf{96} \tag{27}$$

$$\mathbf{97} \xrightarrow[\text{MeOH/dioxane} \atop (37\%)]{e\ (C\ rod),\ Et_4NOTs} \mathbf{98} \tag{28}$$

$$\mathbf{99} \xrightarrow[\text{MeOH/dioxane} \atop (69\%)]{e\ (C\ rod),\ Et_4NOTs} \mathbf{100} \tag{29}$$

$$\mathbf{101a\text{-}c} \xrightarrow[\text{Et}_4\text{NOTs, DMF}]{-2.7\ V\ (SCE)} \mathbf{102a\text{-}c} \quad \begin{array}{l} R = CH_2=CHCH_2\text{-}\ (36\%) \\ R = n\text{-}C_3H_7\ (34\%) \\ R = n\text{-}C_4H_9\ (26\%) \end{array} \tag{30}$$

The reactions are conducted using a very simple setup, consisting of carbon rod electrodes with either a mixed solvent system of methanol/dioxane in an undivided cell, or DMF using a diaphragm; Et$_4$NOTs served as the supporting electrolyte. An initial cathode potential of -2.8 V is used, though the reactions are generally conducted at a constant current of 200 mA.

The selection of Et$_4$NOTs as the supporting electrolyte occasionally presents problems when DMF is used as the solvent. For example, the reduction of keto olefin **103** affords both the cyclized adduct **104**, as well as alcohol **105**. The latter is presumably formed by protonation of the initially formed ketyl, followed by the addition of a second electron, and nucleophilic attack of the resulting carbanion on the quaternary ammonium salt.

$$\text{103} \xrightarrow[\text{Et}_4\text{NOTs, DMF} \atop (26\% \ 104, 37\% \ 105)]{-2.7 \text{ V (SCE)}} \text{104} + \text{105} \qquad (31)$$

Cyclization can also be effected using the solid amalgam, $DMP(Hg)_5$, that deposits at the electrode surface when N,N-dimethylpyrrolidinium perchlorate (DMP^+) is reduced at either a mercury or a lead cathode [48, 49].

$$\text{95} \xrightarrow[\text{DMF (90-94\%)}]{\text{Hg, } \ NMe_2ClO_4} \text{96}$$

$$\underset{DMP^+}{\bigcirc NMe_2ClO_4} + 1\,e + 5Hg \rightleftharpoons DMP(Hg)_5 \qquad (32)$$

$$\downarrow 95$$

$$\text{106} + DMP^+ + 5Hg$$

The amalgam is capable of reducing a number of organic molecules that are otherwise very difficult to reduce [50]. For example, while the reduction potential for fluorobenzene is -2.97 V (SCE), fluoroaromatics can be reduced in the presence of the amalgam, $DMP(Hg)_5$. The latter, is produced via the reduction of DMP^+ at -2.75 V (SCE), i.e., at a value significantly more positive than that required to reduce fluorobenzene. In the presence of the fluoroaromatic **107**, one observes cyclization onto the pendant alkene to afford 1-methylindane (**108**). Furthermore, once the amalgam has delivered an electron, DMP^+ is reformed and therefore serves as a catalyst in this and other transformations.

$$\text{107} \xrightarrow[-2.75 \text{ V, 2 F/mol}]{DMP^+ \ (1 \text{ mmol/l}),} \text{108 (47\%)} + \text{109 (11\%)} \qquad (33)$$

The major product of the keto olefin cyclizations often corresponds to what one would predict, assuming the intermediate ketyl behaves like the corresponding monoradical (Eq. 34). For example, given an option between a 5-*exo* and a 6-*endo-trig* cyclization, the former predominates in radical cyclizations [51], and constitutes the exclusive cyclization path in the electrochemical counterpart, **110** → **112** [47]. In addition, the stereochemical outcome parallels that of the radical process, there being a distinct preference for the formation of cis-1,2-dialkyl substituted alcohols, **112**.

$$R\text{—}CH=CH\text{—}\cdots\text{—}C(=O)R' \; (\mathbf{110}) \xrightarrow{5\text{-}exo\text{-}trig} [\mathbf{111}]^{-\bullet} \longrightarrow \mathbf{112} \quad (34)$$

The principle side reaction corresponds to reduction of the carbonyl without cyclization. For example, reduction of 6-methylhept-6-en-2-one (**113**) leads to a 12% yield of alcohol **114**; no cyclized adduct **115** is produced. Were the intermediate to behave precisely like the monoradical, one would have anticipated that the presence of the methyl group on the alkene would have slowed the rate of 5-*exo-trig* cyclization to a point where closure to form **115**, the product of a 6-*endo-trig* cyclization, would have dominated; it did not.

$$\mathbf{113} \xrightarrow[\text{MeOH/dioxane}]{e\;(C\;\text{rod}),\;Et_4NOTs} \mathbf{114} \quad (12\%) \quad\not\to\quad \mathbf{115} \tag{35}$$

In several instances a comparison was made between the results obtained electrochemically and those obtained using other reducing agents [47]. The results portrayed in Table 6 are striking insofar as the electrochemical protocol proved superior to the alternatives in each instance.

$$\mathbf{116} \xrightarrow{\text{see Table 6}} \mathbf{117} + \mathbf{118} + \mathbf{119} \tag{36}$$

Table 6. Reducing agents compared

Reducing agent	Solvent	Yield (%)		
		117	118	119
Cathode (C rod)	MeOH/dioxane	75	0	0
	DMF	77	0	0
Al(Hg)	PhH	6	0	19
Na	wet Et$_2$O	0	0	65
Na	NH$_3$/THF	6	1	71
Na	HMPA/THF	63	15	0
TiCl$_4$-Mg(Hg)	THF	0	59	0

3.3 Alkene/Ester

Esters are difficult to reduce, and are inert to many of the conditions used in electroreductive processes. A recent investigation has demonstrated that they can easily be reduced at a magnesium cathode in the presence of t-BuOH [52, 53]. When tethered to an alkene, cyclization occurs to afford a cyclic alcohol. Two examples are illustrated, the second being a key step in a synthesis of racemic muscone [53].

$$120 \xrightarrow[(60\%)]{\text{Mg/Mg, } t\text{-BuOH}} 121 \tag{37}$$

$$122 \xrightarrow[(63\%)]{\text{Mg/Mg, } t\text{-BuOH}} 123 \tag{38}$$

3.4 Allene and Alkyne Acceptors/Carbonyl

One of the disadvantages of the chemistry discussed in the previous two sections is that the conversions transform a substance having two or more functional groups, to an adduct possessing one fewer. An interesting and useful variant calls for replacement of the alkene with either an alkyne **124** or an allene **125** [54]. Though modified from its original form, functionality is maintained in the adduct **127**.

$$125 \xrightarrow{\substack{2\text{ e,} \\ 2\text{ HD}}} [126]^* \longrightarrow 127 \longleftarrow [128]^* \xleftarrow{\substack{2\text{ e,} \\ 2\text{ HD}}} 124 \tag{39}$$

Two marine natural products, capnellene diol (**129**) and isoamijiol (**130**), were viewed as ideal candidates upon which to explore these ideas [55]. As illustrated, both of these materials incorporate a bridgehead hydroxyl group and an exocyclic carbon-carbon π-bond, subunits which are produced as a natural consequence of the methodology.

Unfortunately, efforts were thwarted by the formation of the endocyclic alkene **134**. In these cases, use of sodium naphthalide afforded the desired adducts, **135** and **137**, though the yields in both of the cases portrayed in Scheme 8, were poor.

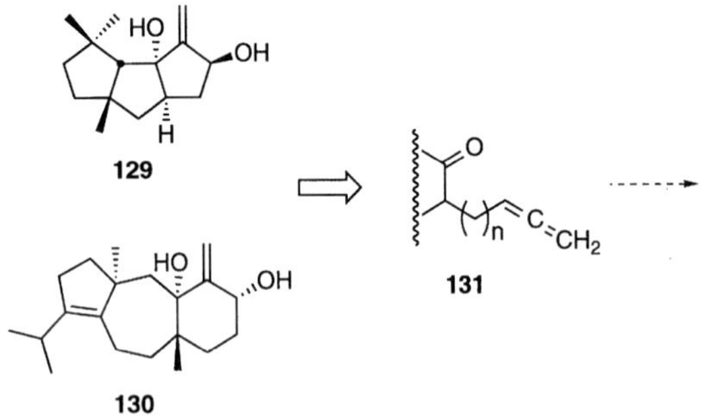

Scheme 8

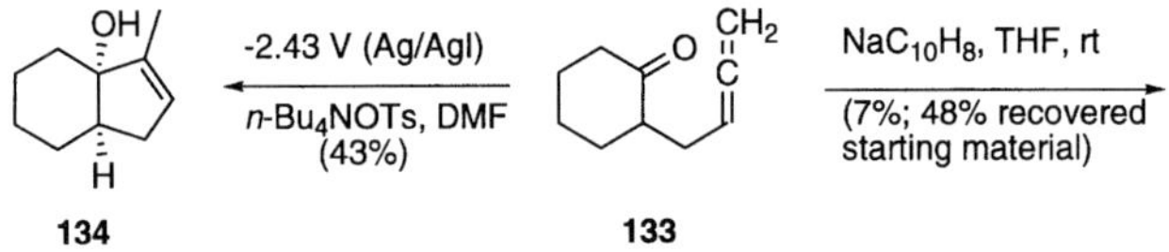

(40)

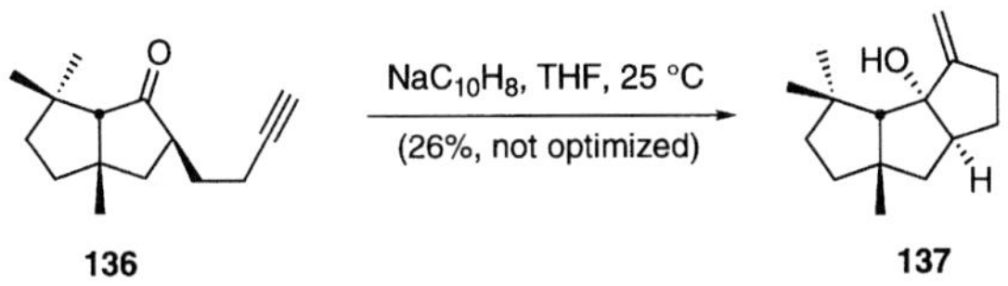

(41)

3.5 Nitrile/Carbonyl

Keto nitriles, such as **138**, function admirably as substrates in reductive cyclizations [56, 57]. Two product types are obtained, one the simple ketone **140**, the other **139**, incorporating the α-hydroxy ketone (ketol) functionality that is present in many natural products (note Eq. 42). Both controlled potential and constant current conditions have been utilized. Of the electrodes examined (Ag, Cd, Pb, Zn, C-fiber, and Sn), tin generally proved most effective. Using tin, the controlled potential reduction of **138** in i-PrOH at -2.8 V vs SCE (divided cell, ceramic diaphragm) afforded a 76% yield of ketol **139** accompanied by 2% of ketone **140**. As illustrated in Table 7, the preference for ketol formation drops when the transformation is carried out at constant current or without using a diaphragm.

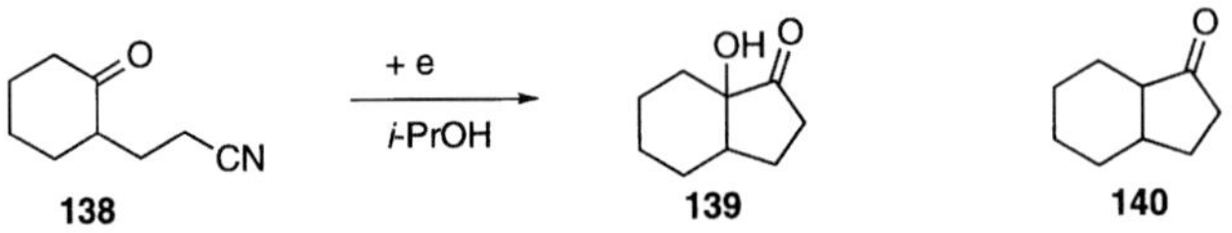

(42)

Table 7. Reduction of keto nitriles

Conditions	F/mol	Yield (% **139**, % **140**)
i-PrOH, 25 °C, −2.8 V (SCE; Sn) divided cell, ceramic diaphragm	3	76, 2
iPrOH, 25 °C, constant current of 0.2 A, divided cell, ceramic diaphragm	3.5	63, 16
iPrOH, 25 °C, −2.8 V (SCE; Sn) divided cell, no diaphragm	4	65, 11

The reaction is assumed to occur via the formation and cyclization of a ketyl, **141**, followed by the addition of a second electron and two protons, leading to an unstable α-hydroxyimine **142**. Hydrolysis affords the ketol **139**, while elimination of water, followed by a second reduction of the resulting α,β-unsaturated imine and hydrolysis, leads to the ketone **140**.

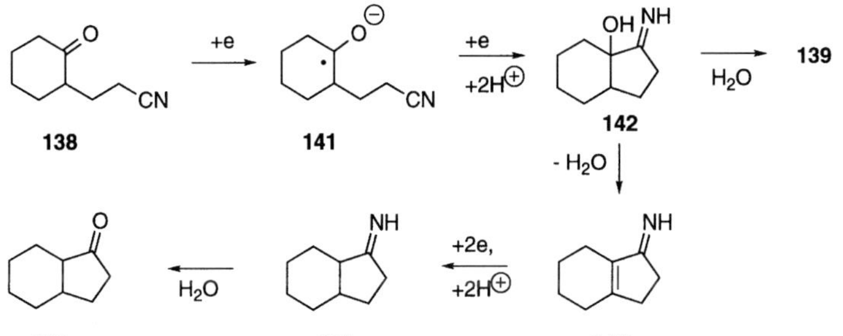

Scheme 9

The transformation has been employed extensively in the preparation of precursors to a number of natural products (note Schemes 10–15), including, guaiazulene (**148**), dihydrojasmone (**150**), rosaprostol (**155**), and convergent syntheses of (−)-valeranone (**159**), hirsutene (**162**), and $\Delta^{9(12)}$-capnellene (**167**) [57]. The key cathodic cyclization step is illustrated for each natural product.

α-Ketols can also be synthesized using a novel, electrochemically generated acyl anion equivalent [58–61]. For example, the constant current electrolysis of a δ-keto carboxylic acid in the presence of tributylphosphine and methanesulfonic acid at 0 °C in an undivided cell, affords an α-ketol in yields ranging from 16–57%; note Eqs. (43) and (44). Presumably, the reaction proceeds through the intermediate phosphonium salts **169** and **170**, the former being formed at the anode. Reduction and protonation of **170** leads to ylide **172** which subsequently undergoes cyclization to afford **173**. Given the simplicity and accessibility of the starting materials, this methodology provides a very convenient route to ketols.

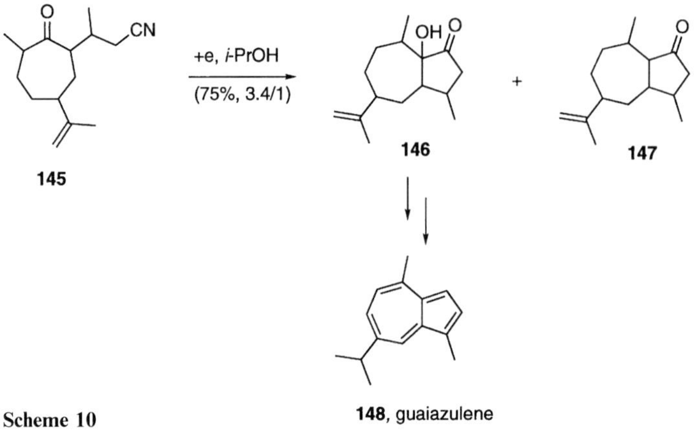

Scheme 10

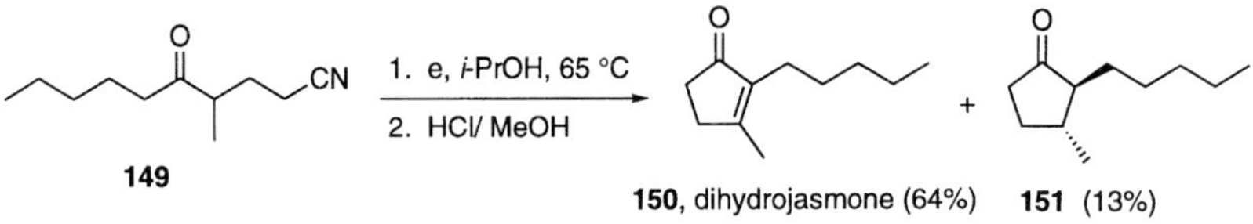

146, 147, 148, guaiazulene

Scheme 11

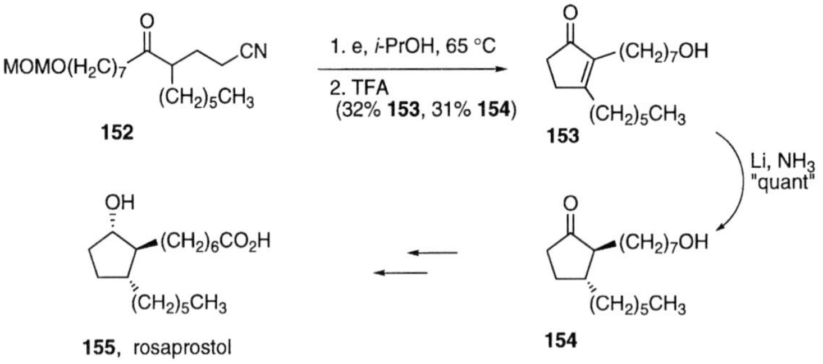

149 → 150, dihydrojasmone (64%) + 151 (13%)

Scheme 12

152 → 153 (32%), 154 (31%) → 155, rosaprostol

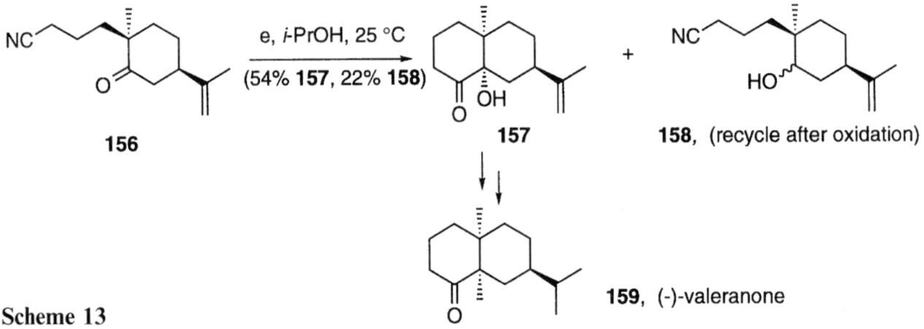

Scheme 13

156 → 157 (54%) + 158 (recycle after oxidation) → 159, (−)-valeranone

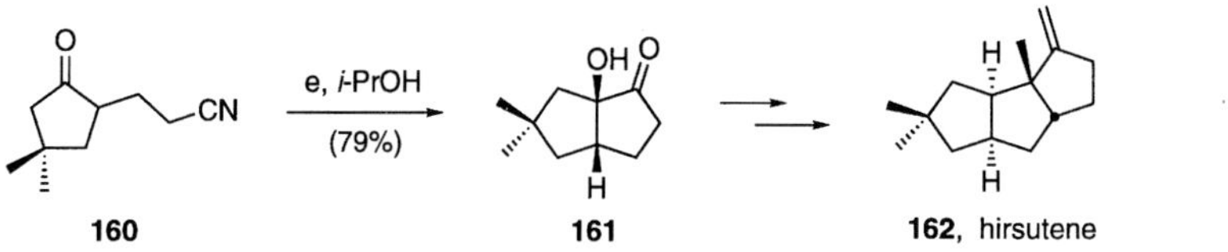

Scheme 14

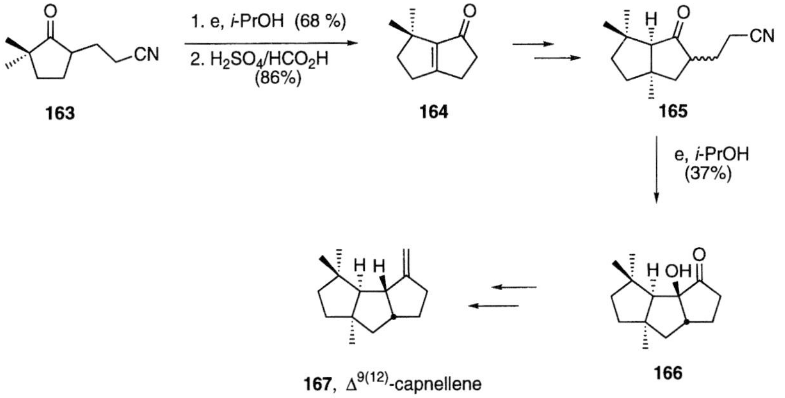

Scheme 15

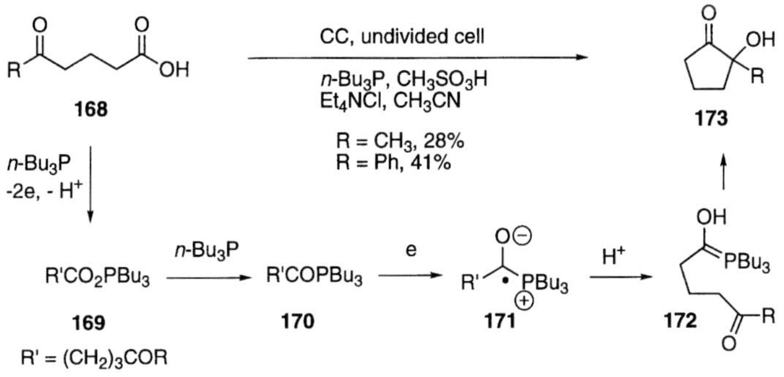

(43)

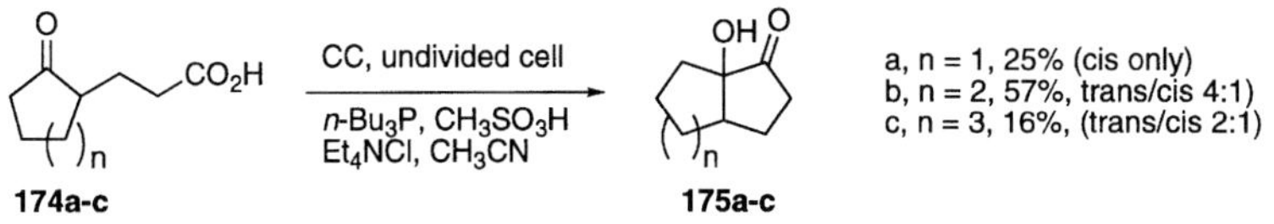

(44)

The process can be interrupted at the stage of ketyl **171**, thereby providing access to an acyl radical synthon. The radical character of the intermediate is demonstrated by the closure onto pendant alkenes (Table 8). Unfortunately, cyclization fails unless the terminal carbon of the alkene is substituted with an aryl group. Clearly, these intermediates are not as reactive as other acyl radical equivalents.

$$\underset{\textbf{176a-f}}{R\diagdown\diagup\diagdown CO_2H} \xrightarrow[n\text{-Bu}_4NBr,\ CH_2Cl_2,\ rt]{CC,\ n\text{-Bu}_3P,\ CH_3SO_3H} \underset{\textbf{177a-f}}{\text{cyclopentanone}} + \underset{\textbf{178a-f}}{R\diagdown\diagup\diagdown CHO} \quad (45)$$

Table 8. Cyclization of carboxylic acids onto alkenes: acyl radical synthon

	R	R'	Yield 177 (%)	Yield 178 (%)
a,	Ph	H	59	37
b,	$(CH_2)_2Ph$	H	68	–
c,	p-ClC$_6$H$_4$	H	45	44
d,	p-BrC$_6$H$_4$	H	42	38
e,	p-MeOC$_6$H$_4$	H	62	20
f,	Ph	Me	82	9

3.6 The Aromatic/Carbonyl Variation

One of the more remarkable reductive cyclization processes involves closure of a ketyl onto an aromatic ring, a transformation which would appear to be thermodynamically unfavorable [62]. The factors influencing the course of the reaction have been studied in some detail. A tin cathode is preferred, as are i-PrOH and Et$_4$NOTs as the solvent and supporting electrolyte, respectively. Inspection of the structure of a typical product such as **180** reveals that, if **180** aromatized, the process accomplishes essentially the same thing as a two-step sequence involving a Friedel-Crafts acylation, followed by the addition of an organometallic to the resulting ketone. The principle side reaction is the simple reduction of the ketone to the corresponding alcohol, **181**.

$$\underset{\textbf{179}}{\text{PhCH}_2)_3COCH_3} \xrightarrow[\text{4 F/mol (70\% 180, 7\% 181)}]{e\ (Sn/C),\ i\text{-PrOH, Et}_4NOTs,} \underset{\textbf{180}}{\text{cis-decalinol}} + \underset{\textbf{181}}{\text{Ph(CH}_2)_3CH(OH)Me} \quad (46)$$

As illustrated, the transformation is stereoselective, displaying a significant preference for formation of the product wherein the ring junction hydrogen and hydroxyl group on the vicinal carbon are *cis* to one another. Furthermore, in those instances where there are substituents on the tether (e.g., **182a,b**), cyclization occurs in a manner wherein they occupy equatorial orientations about the ring being formed **183a, b**. Unfortunately, cyclization fails to occur when hindered ketones are used, and in attempts to form ring sizes other than six; note Eqs. (48) and (49).

$$\text{182a, R = H, R' = CH}_3 \text{ (68\%)} \quad \xrightarrow{\text{e, } i\text{-PrOH}} \quad \textbf{183a, 183b} \tag{47}$$
$$\textbf{182b, R = CH}_3\text{, R' = H (60\%)}$$

$$\textbf{184} \quad \xrightarrow[\text{(88\%)}]{\text{e, } i\text{-PrOH}} \quad \textbf{185} \tag{48}$$

$$\textbf{186} \quad \xrightarrow[\text{(80-90\%)}]{\text{e, } i\text{-PrOH}} \quad \textbf{187} \tag{49}$$

As illustrated in the accompanying equations, the regiochemical outcome is influenced by the nature of the substituents appended to the aromatic ring, it dropping significantly when an electron-donating group is appended *meta* to the tether [62]. This is not surprising given the significant electron density at that carbon in the putative intermediate **195** portrayed in Scheme 16; an electron-donating group should discourage its formation.

$$\textbf{188} \quad \xrightarrow{\text{e, } i\text{-PrOH}} \quad \textbf{189 (17\%)} \quad + \text{ 36\% uncyclized alcohol} \tag{50}$$

$$\textbf{190} \quad \xrightarrow[\substack{\text{2. H}_2\text{, Pd/C} \\ \text{(62\%, two steps)}}]{\text{1. e, } i\text{-PrOH}} \quad \textbf{191} \quad + \quad \textbf{192} \tag{51}$$

The electrochemically promoted cyclization has proven far superior to the use of traditional reducing agents [62]. Table 9 illustrates the range of reducing agents and conditions that have been examined. With the exception of sodium in 2:1 HMPA/THF, they all failed to afford cyclized material.

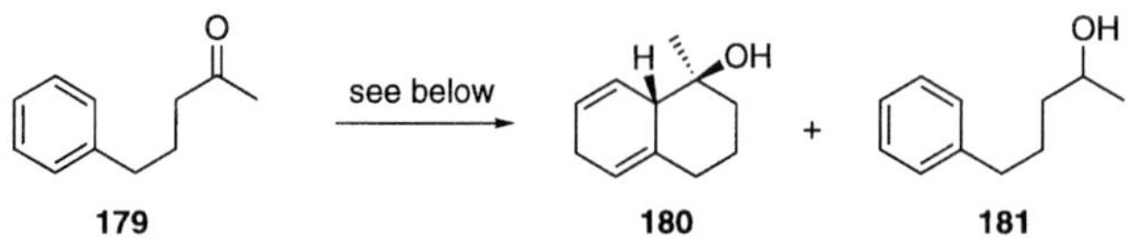

(52)

Table 9. Ketyl cyclizations onto aromatics

Reducing agent	Conditions	% 180	% 181
cathode (Sn)	Et$_4$NOTs, i-PrOH, 25 °C	70	7
Zn	NaOH, i-PrOH, 65 °C	0	84
Sn	NaOH, i-PrOH, 65 °C	0	85
Na	i-PrOH, 25 °C	0	81
Na	wet Et$_2$O, 25 °C	0	90
Na	NH$_3$/THF, −70 °C	0	0
TiCl$_4$-Zn	THF, 65 °C	0	35
SmI$_2$	t-BuOH/HMPA/THF, 0 °C	0	trace
Na	HMPA/THF (2:1), 0 °C	42	17

The process is believed to occur via an initial reduction of the ketone, followed by cyclization of the resulting ketyl, **193a (193b)**. The stereochemical preference is rationalized by suggesting that repulsion between the negative charge on oxygen and the π-cloud of the aromatic ring, as in structure **193b**, discourages cyclization to **196**, the precursor to an adduct wherein the hydroxyl group and ring junction hydrogen are *trans* to one another.

In principle, conversion of radical anion **195** to the product **180** could occur via abstraction of a hydrogen atom and then a proton from the solvent (path A, Scheme 17), or via reduction to afford a dianion which subsequently abstracts two protons from the solvent (path B) [62]. To differentiate between these alternatives, the reduction of **179** was carried out in *i*-PrOD. Were pathway

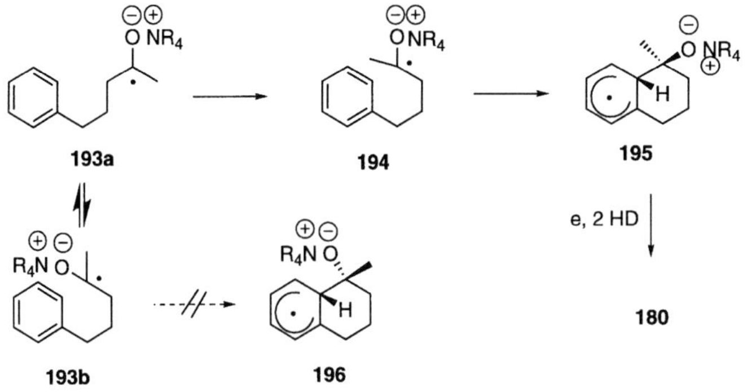

Scheme 16

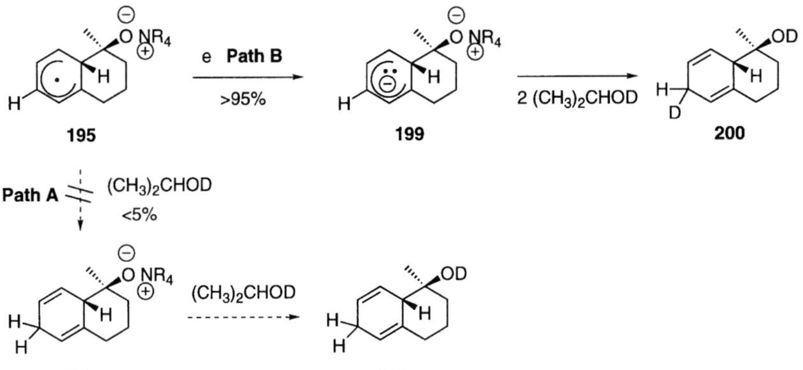

Scheme 17

A operable, one would anticipate hydrogen atom incorporation into the aromatic ring to afford **198**, while the alternative pathway places deuterium there and would lead to compound **200**. The data indicates a clear and distinct preference for the latter (> 95% **200**).

One final point of interest concerns the remarkable influence of counterion. Cyclization fails, for example, when lithium perchlorate is used in place of a tetraalkylammonium salt as the supporting electrolyte. The rationale for this interesting observation focuses upon the ease with which the initially formed intermediate **201** is expected to undergo further reduction as a function of the counterion, M. When it is a quaternary ammonium ion as in **201b**, the addition of a second electron ought to be slower than when M=Li (**201a**), given that the first species possesses substantial ionic character, while in the alternative, the O–Li bond is covalent. Thus, with a lithium counterion, the addition of a second electron ought to occur more rapidly and be followed by protonation rather than cyclization; this is the case.

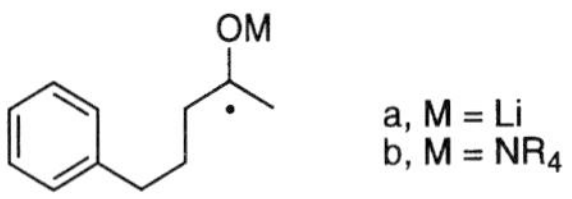

3.7 Pyridinium/Carbonyl

Pyridinium salts tethered to ketones also undergo cathodic cyclization [1]. The reaction provides a convenient diastereoselective route to quinolizidine and indolizidine derivatives such as **203**, **204** and **206**, **208**, and **209**, and appears to hold significant promise as a route to alkaloids. Examples are portrayed and the optimal conditions are listed below the equations. A mercury cathode is preferred, as passivation occurs when lead is used, and the reaction does not occur using either graphite or glassy carbon electrodes.

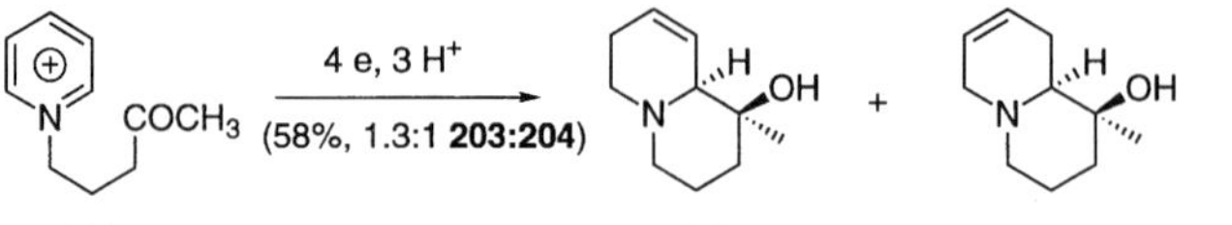

$$\text{202} \xrightarrow{\text{4 e, 3 H}^+}_{(58\%,\ 1.3:1\ \textbf{203}:\textbf{204})} \textbf{203} + \textbf{204} \tag{53}$$

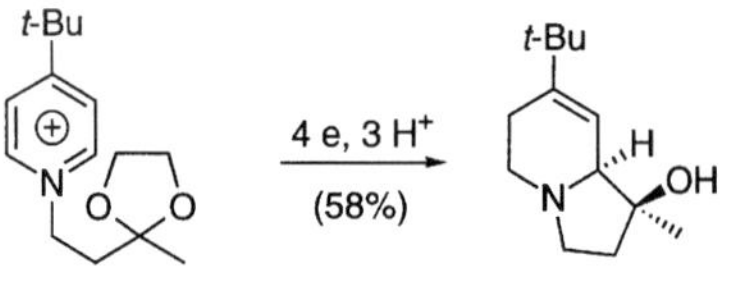

$$\textbf{205} \xrightarrow{\text{4 e, 3 H}^+}_{(58\%)} \textbf{206} \tag{54}$$

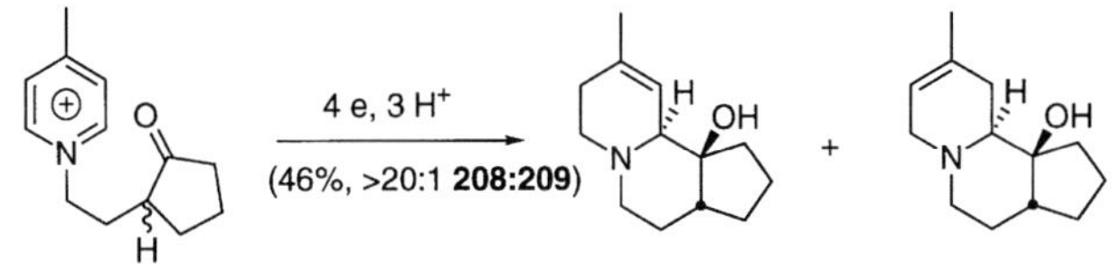

$$\textbf{207} \xrightarrow{\text{4 e, 3 H}^+}_{(46\%,\ >20:1\ \textbf{208}:\textbf{209})} \textbf{208} + \textbf{209} \tag{55}$$

Hg (constant current of 4.2 mA cm^{-2}, charge 8 F mol^{-1}), 10% H_2SO_4, 0.2 M in pyridinium salt, 20 °C)

In each instance, the product corresponds to that diastereomer wherein the hydroxyl group and ring junction hydrogen are *trans* to one another. AM1 calculations place a partial negative charge on nitrogen in the radical, **211**, formed after the initial protonation and subsequent reduction of the carbonyl [1]. This information lends credibility to the existence of a hydrogen bond, as shown in structures **211** and **212**, that positions the radical so that attack on the

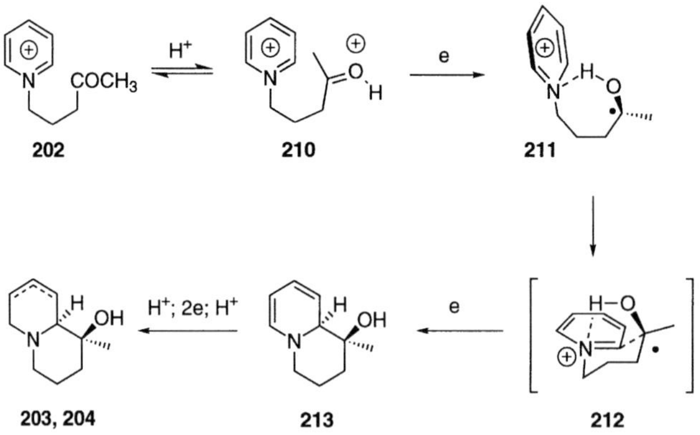

Scheme 18

side of the pyridinium ring which is proximal to the H-bond leads to the observed stereochemical outcome.

4 Tandem Cyclizations

Given the large number of tandem cyclization processes that have been explored [63], it is disappointing to note that so few have been promoted electrochemically. There appears to be a significant opportunity for additional exploration. Two types of tandem cathodic cyclizations are discussed below. The first involves generation of a ketyl, and its subsequent cyclization onto a pendant alkene to afford a new radical that closes onto a second alkene [64, 65]. The second focuses on chemistry not yet discussed involving the reductive cyclization of enol phosphates of 1,3-dicarbonyl compounds [66].

4.1 Tandem Cyclization of Non-Conjugated Dienones

The controlled potential cathodic reduction of the non-conjugated dienone **214** affords the bicyclic alcohol **215** in a 54% yield [64]. The process has also been examined under both controlled potential (CP) and constant current (CC) conditions, the latter proving advantageous for reactions involving the use of a gram or more of substrate. Under CC conditions, a "small amount" of **216**, the adduct resulting from one rather than two cyclizations, is produced. The reaction does not always proceed with high stereoselectivity. This is evidenced, for example, by the production of the epimeric ethers **218a, b** in the tandem cyclization of enol ether **217**.

Unfortunately, the scope of the process appears to be limited by the need to form a five-membered ring in the second ring closure event [64]. For example, reduction of **220** affords **221** and **222**, each the result of a 5-*exo-trig* cyclization (Table 10). Apparently, the rate of 6-*exo-trig* cyclization of the putative intermediate **223** is slower than the addition of a second electron and protonation; no bicyclic adduct is produced [65].

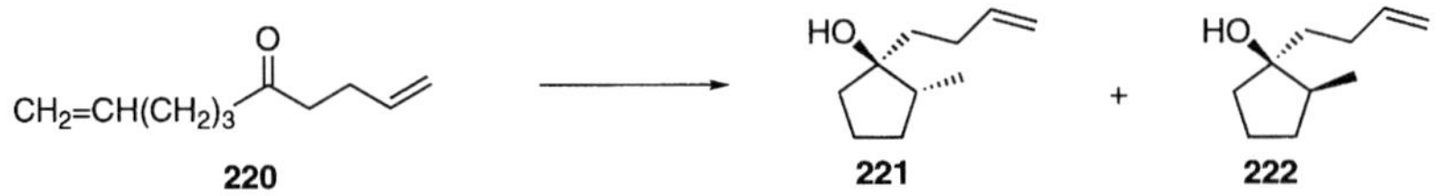

(58)

Table 10. Tandem cyclizations of keto olefins

Conditions	Yield **221** (%)	Yield **222** (%)
e (Hg/Pt), DMF, CC 0.1 mol/l n-Bu$_4$NBF$_4$	60	40
e (Hg/Pt), DMF, CC 0.1 mol/l n-Bu$_4$NBF$_4$, 0.18 mol/l i-PrOH	87	13
e (Hg/Pt), DMF, CC 0.1 mol/l n-Bu$_4$NBF$_4$, 5 mmol/l DMPBF$_4$	50	31

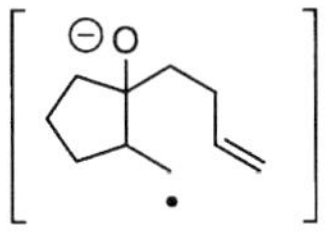

On the other hand, when the order of events is reversed, as is the case with **224**, so that cyclization to a six-membered ring is followed by a more rapid 5-*exo-trig* cyclization, then the bicyclic framework **225** is produced, though the yield is not high. This is unfortunate, given the simplicity of the starting material and the potential for transforming that simplicity into a more complex structure in a single step.

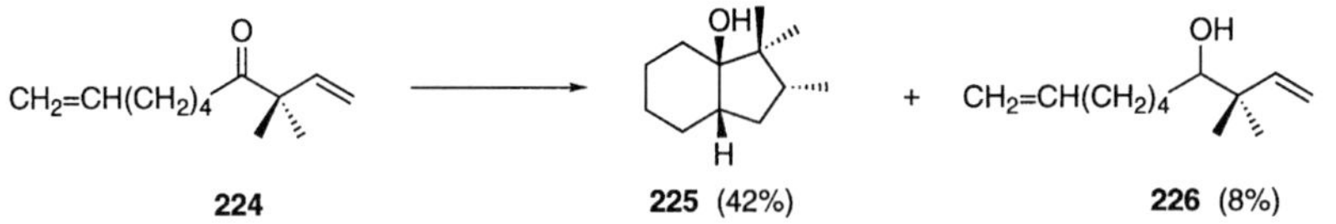

(59)

constant current (6.6 mA/cm^2 to maintain -2.66 V), DMF, 0.1 mol/l n-Bu$_4$NBF$_4$

4.2 Enol Phosphates of 1,3-Dicarbonyl Compounds

The reductive cyclization of readily available enol phosphates of 1,3-dicarbonyl compounds bearing pendant olefinic units has been explored [66, 67]. The chemistry is exceptionally interesting, and provides a unique route to structures possessing a cyclopropyl unit which is suitable for structural elaboration. The reaction occurs in a manner wherein the phosphate-bearing carbon behaves like a carbene that adds to the pendant alkene to form a cyclopropane. While this provides a useful way of viewing the transformation, mechanistic studies indicate that a carbene is not an actual intermediate. Examples are portrayed in Table 11.

Table 11. Reductive cyclization of enol phosphates

Starting material	Product[a]	Yield (%)
227	228	58
229	230	38
231	232	63
233	234, R = CO_2Et, R' = H 235, R = H, R' = CO_2Et	26 22

[a] e (Hg), 0.2 mol/l n-Bu$_4$NClO$_4$, DMF, -2.05 V (SCE), add substrate to maintain a current between 100 and 130 mA. After substrate has been added, increase to -2.30 V and maintain there until current drops below 10 mA

The initial stage of the process undoubtedly involves reduction of the unsaturated ester, followed by extrusion of phosphate. Details concerning the latter stages remain obscure. Evidence has been obtained to rule out the intermediacy of a singlet carbene. To do so, the chemistry of the E/Z-isomers **236** and **237** was investigated [67]. In each instance, a mixture of stereoisomeric products was obtained from a geometrically pure isomer of the starting material. The attendant loss of stereochemistry argues against a singlet carbene. Notice, however, that the product isomer ratio, **238/239** or **240/241**, differs depending upon which isomer of the starting material was reduced. From this information, one can conclude that, regardless of the precise nature of the intermediate, it must not live long enough to allow complete equilibration to occur.

236 → e (Hg), 0.2 mol/l n-Bu$_4$NClO$_4$, DMF, -2.05 V (initially) to -2.30 V

238, 51%, R = H, R' = CH_3
239, 10%, R = CH_3, R' = H

(60)

$$\text{237} \xrightarrow{\text{e (Hg), 0.2 mol/l } n\text{-Bu}_4\text{NClO}_4, \text{ DMF, -2.05 V (initially) to -2.30 V}} \text{240, 41\%, R = H, R' = CH}_3 \quad \text{241, 18\%, R = CH}_3\text{, R' = H} \tag{61}$$

Given this information, it was suggested that cyclization may occur either via the intermediacy of a vinyl radical **242** or the corresponding carbanion **243**. However, no evidence is available to allow differentiation between these options.

$$\text{242, 243} \longrightarrow \text{244} \longrightarrow \text{245} \tag{62}$$

* = • or ⊖

4.3 Tandem Cyclization of Enol Phosphates

The opportunity for tandem cyclization was explored. Here, the results can be accommodated by postulating the intermediacy of a vinyl radical [66]. For example, the controlled potential reduction of enol phosphate **246** affords **247** as a mixture of stereoisomers, in addition to a 15% yield of the linearly fused tricyclopentanoid **248**. Assuming that the initial reduction cleaves the phosphate unit, then there exists the opportunity for the resulting radical **249** to be further reduced to afford a carbanion, or undergo a 5-*exo-trig* radical cyclization onto the pendant alkene. Given the nature of the products and the fact that they are inconsistent with the expectations of carbanion chemistry, it seems clear that the latter pathway dominates.

$$\text{246} \xrightarrow{\text{e (Hg; -2.3 V), } n\text{-Bu}_4\text{NClO}_4, \text{ DMF}} \text{247, (53\%, exo/endo 3/1)} + \text{248, (15\%)} \tag{63}$$

One cannot help but wonder whether this chemistry could be exploited in the assembly of natural products. For example, could the yield for formation of the linearly fused tricyclopentanoid **248**, a structural unit common to many naturally occurring materials, be increased through the systematic variation of reaction conditions?

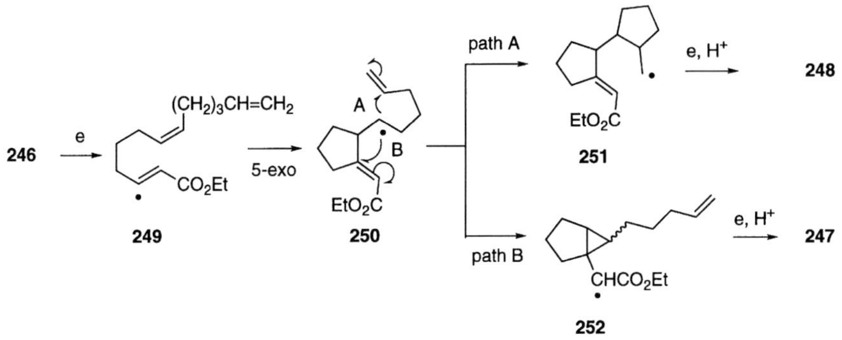

Scheme 19

5 Displacements

The use of lithium in liquid ammonia to reduce enones is a well-known, well-established procedure which has seen widespread use. The nucleophilic character of the β-carbon is clear, and has been demonstrated in many ways. For example, reduction of enone **253** leads to displacement of tosylate and formation of the tricyclic ketone **254** [68, 69]. It is interesting to note that the yield for formation of **254** is a function of the nature of the reducing agent. For example, using Li/NH_3, a 45% yield is obtained, while with lithium dimethylcuprate, it is 96% [70], and via cathodic reduction, 98%.

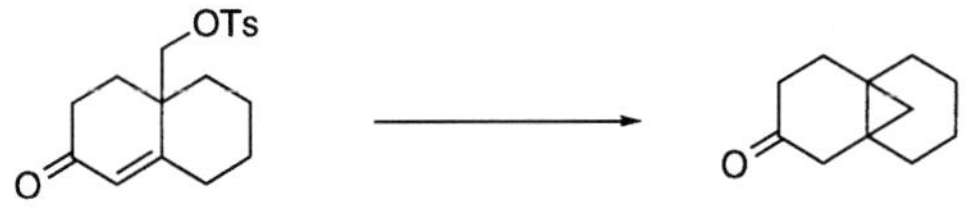

reagents	yield (%)
Li/NH_3	45
$LiCuMe_2$	96
cathode	98

(64)

In principle, the reaction might have occurred via an alternative pathway involving the initial reductive cleavage of the tosylate to form a carbanion which subsequently undergoes an intramolecular Michael reaction [71]. To shed light on the mechanism, and to expand the scope of the process, the mesylates portrayed in Table 12 were synthesized and examined. The following points are noteworthy. (a) The yield of cyclized material increases in an order that roughly parallels the relative rates of intramolecular displacement. That is, the rate of cyclization leading to a given ring size, $k_r(n)$, where n = size of the ring, is

Table 12. Yields of cyclized products vs ring size

	n	Ring size	Working potential (V)	Yield (%) in the absence of $CH_2(CO_2Me)_2$	Yield (%) in the presence of $CH_2(CO_2Me)_2$
a,	1	4	−2.3	6	–
b,	2	5	−2.2	83	46
c,	3	6	−2.3	58	–
d,	4	7	−2.3	4	–

$k_r(3) \gg k_r(5) \geq k_r(6) > k_r(7) > k_r(4)$. (b) The addition of dimethyl malonate as a proton donor reduces the amount of cyclization leading to the five-membered ring, has no effect on the formation of the cyclopropyl ester, and eliminates cyclization leading to the four-, six-, and seven-membered rings. These results clearly suggest that, under the reaction conditions, the rate of proton transfer exceeds that of cyclization for formation of the latter three ring sizes, but is not competitive with the rate for formation of the three-membered ring. (c) The peak potentials, E_p, for the reduction of ethyl crotonate as well as compounds **257a,c,d** are approximately the same, ranging from −2.47 to −2.49 V, as determined using cyclic voltammetry at a scan rate of 100 mV/sec. On the other hand, E_p for the three- and five-membered ring precursors, **255** and **257b**, is shifted to more positive values, −2.32 V and −2.41 V, respectively.

$$MsO\text{–}CH(CH_3)\text{–}CH=CH\text{–}CO_2Et \xrightarrow{-2.2\ V} \triangleright\text{–}CH_2CO_2Et \quad (65)$$

255 → **256**
91% (in the absence of malonate)
88% (in the presence of malonate)

$$MsO\text{–}(CH_2)_n\text{–}CH=CH\text{–}CO_2Et \xrightarrow{e^-} \text{(cyclic)}_n\text{–}CH_2CO_2Et \quad (66)$$

257a-d → **258a-d**

These observations are in accord with a scheme involving a reversible electron transfer, followed by a reaction that depletes the concentration of the initially formed reduced species, R. They are also reminiscent of the observations made earlier in regard to the electrohydrocyclization process. The greater the rate of the follow-up process, the more significant its effect on the concentration of R in a given time period, that associated with the CV scan rate, for example. From a moments consideration of the Nernst equation, it is clear that this event should manifest itself in terms of a shift in the peak potential to a more positive value, as observed for **255** and **257b** [4]. In the present instance, it is suggested that a rapid or concerted loss of the mesylate anion in the reductive cyclization is likely to be associated with this so-called "kinetic shift" of the peak potentials [69].

$$O \underset{-e}{\overset{+e}{\rightleftarrows}} R \xrightarrow{k} P \quad (67)$$

The issue of whether cyclization occurs via reduction of the unsaturated ester, followed by cyclization with displacement of the mesylate, or instead via an initial reductive cleavage of the C–O bond followed by an intramolecular conjugate addition, was ascertained in a clever manner using a stereochemical probe [69]. Thus, when the isomeric mesylates **259** and **260** were subjected to the same conditions, the *cis*-isomer **260** failed to undergo cyclization, while its *trans* counterpart, **259**, afforded a 70% yield of stereoisomeric products, **261** and **262**. Were the process to involve an initial cleavage of the mesylate, then both starting materials would have afforded the same intermediate and, consequently, the same products. That this was not the case supports the notion that the enoate is reduced first. The resulting radical anion is able to assume a geometry that permits backside displacement of the mesylate in **259a**, but not with **260a**.

6 Halides

6.1 Cyclizations Involving Alkylidene Malonates

Bond constructions similar to those just discussed can be achieved using an alkylidene malonate which is tethered to an alkyl bromide [72]. Of particular interest in this context is the controlled potential reductive cyclization of **263**. As illustrated, the method provides a reasonably facile and modestly efficient entry to cyclobutanes **264**. Presumably, the process is initiated by reduction of the alkylidene malonate rather than the alkyl halide, since alkyl bromides are more difficult to reduce. The same substrate, when reduced with L-Selectride undergoes conjugate addition of hydride and a subsequent cyclization leading to the five-membered ring **265**. The latter transformation constitutes an example of a MIRC reaction [71–73], a process which is clearly complementary to the electrochemical analog.

$$\text{263} \xrightarrow{\substack{-1.85\text{V, DMF,} \\ n\text{-Bu}_4\text{NBr, rt} \\ (65\text{-}80\%)}} \text{264} \quad (70)$$

$$\text{263} \xrightarrow{\substack{\text{Li}(s\text{-Bu})_3\text{BH, THF, 0 °C} \\ (65\%)}} \text{265} \quad (71)$$

$$\text{266} \xrightarrow{\substack{-1.85\text{V, DMF,} \\ n\text{-Bu}_4\text{NBr, rt} \\ (60\%)}} \text{267} \quad (73)$$

$$\text{266} \xrightarrow{\substack{\text{Li}(s\text{-Bu})_3\text{BH,} \\ \text{THF, reflux} \\ (45\%)}} \text{268} \quad (74)$$

6.2 Vitamin B_{12}-Mediated Cyclizations

Similar transformations can be carried out via indirect electrolysis using vitamin B_{12} and B_{12} analogs as mediators [74]. As shown in the accompanying equations, the methodology lends itself nicely to the formation of both spiro and linearly fused materials (Tables 13 and 14) [75].

$$\text{269a-c} \xrightarrow{\substack{0.05 \text{ mol/l vitamin B}_{12}, \\ 0.1 \text{ N LiClO}_4, 0.05 \text{ N} \\ \text{NH}_4\text{Br, DMF, -1.9V}}} \text{271a-c} + \text{272a-c} \quad (74)$$

Table 13. B_{12}-mediated formation of fused adducts.

	n	Yield (%) **271**	Yield (%) **272**
a,	3	< 2	90
b,	4	95	< 2
c,	5	70	10

The chemistry takes place via an initial reduction of vitamin B_{12} or a similar cobalt (III) species **275**, in a process that sees the conversion of cobalt from the +3 to the +1 oxidation state, and the opening of two sites of unsaturation, to afford **276** [74]. This very reactive, highly nucleophilic intermediate reacts rapidly with the alkyl halide to form the octahedral complex **277**, and reestablish

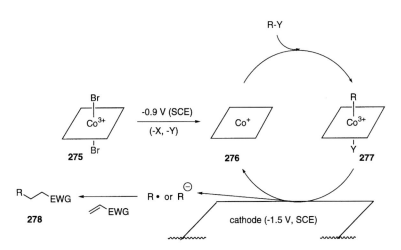

Table 14. B_{12}-mediated formation of spirocycles

	n	Yield (%) 273	Yield (%) 274
a,	3	< 2	90
b,	4	95	< 2
c,	5	45	40

Scheme 20

a +3 oxidation state of cobalt. A catalytic cycle is secured when this material undergoes reaction with a typical Michael acceptor leading to the formation of product **278**.

Control experiments establish that the initial process converting the dibromide **275** to **276** takes place in two steps with $E_{1/2}$ for extrusion of the first and second axial ligands occurring at −0.36 and −1.08 V, respectively [75]. After reaction with the alkyl halide, the resulting octahedral complex **277** is further reduced in the range of −1.4 to −1.7 V to form a cobalt (II) complex which decays via the addition of an additional electron, cleavage of the C–Co and Co–Y bonds, and reaction with the Michael acceptor.

Typically, the reactions are carried out under controlled potential conditions, with the potential being set within the range of −1.4 to −1.9 V (Ag/Ag^+).

This is well shy of that needed to reduce either the enone or alkyl halide. Only 1–20 mol% of the mediator is needed, in a medium consisting of $LiClO_4$ as the supporting electrolyte, DMF as solvent, and NH_4Br as the proton donor.

While chemistry with alkyl halides has been illustrated, it is important to note that **276** also undergoes reaction with epoxides and acid anhydrides [75]. The latter is particularly significant in that it provides an exceptionally simple means of transferring an acyl group to the β-carbon of a typical Michael acceptor. The electrochemically-based methodology appears superior to other methods, occasionally rather elaborate, involving organometallic chemistry. The only obvious drawback involves the use of symmetrical anhydrides, wherein half of the molecule is essentially discarded. This methodology has proved useful in the assembly of natural products, though inter- rather than intramolecular processes have been used in nearly all instances [76, 77].

6.3 Alkynyl Halides

Chloropyridinecobaloxime(III) **279** also serves as a convenient source of nucleophilic cobalt, and as a means of promoting radical cyclizations [77]. For example, its reduction in the presence of a bromoalkyne such as **280** leads to the formation of a radical **281** which closes onto the alkyne, leading eventually to the E,Z-alkenes **283** and **284** in an 81% yield. Cyclizations of this type are obviously complementary to those effected using tributyltin hydride.

In the present instance, the mediated process is more efficient than that promoted by tin (81 vs 73%). Generally, the reactions are conducted with the consumption of ca. 2–5.5 F/mol.

A recent study focused on determination of the minimum amount of cobaloxime needed to serve as an electron carrier [78, 79]. The electrolyses were carried out using a zinc anode and cathode, with variable amounts of cobaloxime. In its absence, a 5% yield of **286** was isolated. The addition of as little

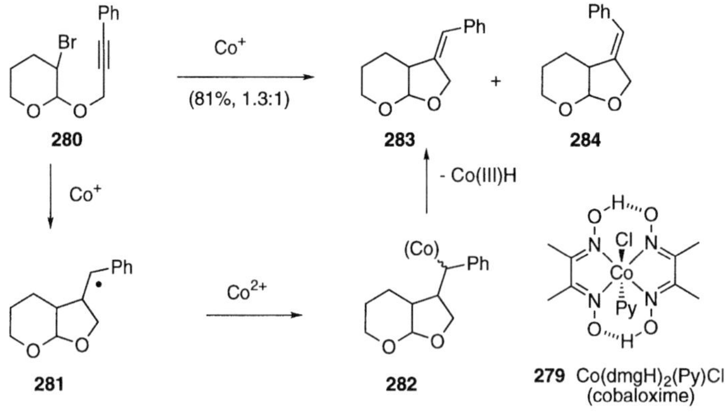

Scheme 21

as 5 mol% afforded an increase in yield to 68%. The yield plateaued at 77%, using 20 mol% of the mediator.

cobaloxime mol %	yield (%)
0	5
5	68
20	77
50	70

(76)

285 → **286**

cobaloxime, Zn/Zn, undivided cell, MeOH, Et$_4$NOTs

The direct reduction of haloalkynes using either mercury or vitreous carbon as the cathode has been examined in considerable detail; [80–84] one example is portrayed in Eq (77). The influence of reduction potential, current consumption, proton donor, electrode, and substrate concentration on the course of the process has been examined. Vitreous carbon electrodes are preferred, though mercury has been used in many instances. Unfortunately, these reactions suffer from the formation of diorganomercurials. While both alkyl iodides and bromides can be used, the former is generally preferred. Because of their higher reduction potential, alkyl chlorides react via a different mechanism, one involving isomerization to an allene followed by cyclization [83].

287 → **288** (60%) + **289** (31%) + **290** (1%)   (77)

C-cathode, DMF, (CF$_3$)$_2$CHOH, R$_4$NClO$_4$

6.4 Allylic and Aryl Halides

Allylic and aryl halides serve as convenient sources of electrogenerated organometallics [85]. The first example portrays a novel route to the cephalosporin derivative **293** [86], it being formed through the intermediacy of the allyl nickel species **292**. The presence of catalytic amounts of NiCl$_2$(bpy) and PbBr$_2$ is essential for the formation of product [85]. The second example is reminiscent of Stille-coupling [87]. The putative carbanion **297** undergoes reaction with electrophiles, El, (e.g., **297** → **298**), in this case carbon dioxide or a proton source, to provide a convenient route to indoles.

291 → **292** → **293**   (78)
*, 53%

R^1 = NHCOCH$_2$Ph
R^2 = CH$_2$C$_6$H$_4$OMe-p * NiCl$_2$(bpy) (0.1 equiv), PbBr$_2$ (0.05 equiv), Pt-Al, CC (6.7 mA/cm^2, 3.23 F/mol)

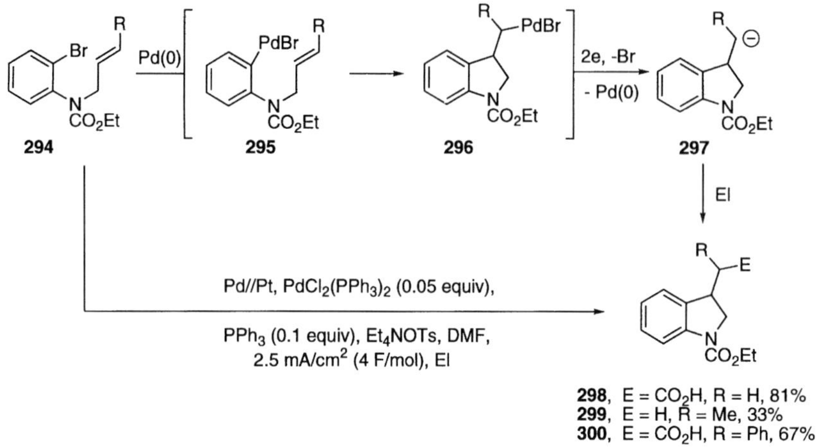

[Scheme showing conversion of 294 → 295 → 296 → 297 → products 298–300 via Pd(0) catalysis]

Pd//Pt, PdCl$_2$(PPh$_3$)$_2$ (0.05 equiv),
PPh$_3$ (0.1 equiv), Et$_4$NOTs, DMF,
2.5 mA/cm^2 (4 F/mol), El

298, E = CO$_2$H, R = H, 81%
299, E = H, R = Me, 33%
300, E = CO$_2$H, R = Ph, 67%

(79)

β-Bromoacetals and β-bromoacetates readily undergo nickel(II) catalyzed cathodic cyclization onto alkenes [88]. The process has its obvious counterpart in free radical initiated chemistry using an organostannane, organomercury reagent, or an organosilane. These methods generally utilize stochiometric quantities of the initiator, and often require the use of high dilution. In contrast, the use of [Ni(cyclam)](CLO$_4$)$_2$ as a mediator avoids high dilution and allows the process to take place in the presence of only 10 mol% of the cyclam. As expected and illustrated in the following examples, cyclization is most effective when an electron-deficient alkene is utilized. Of particular note is the fact that the product of the first reaction, **302**, is an intermediate in the synthesis of *Ipecac* and *Corynanthe* alkaloids, while the lactam, **306**, is a synthetic precursor of the alkaloid tacamonine.

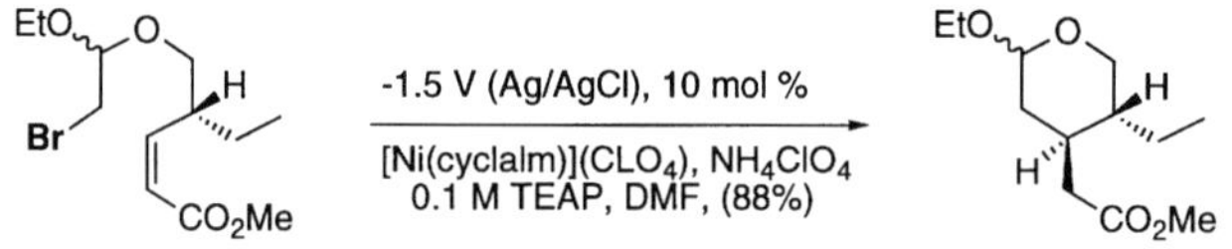

301 →(−1.5 V (Ag/AgCl), 10 mol % [Ni(cyclalm)](CLO$_4$), NH$_4$ClO$_4$, 0.1 M TEAP, DMF, (88%))→ 302 (80)

303 →(−1.5 V (Ag/AgCl), 10 mol % [Ni(cyclalm)](CLO$_4$), NH$_4$ClO$_4$, 0.1 M TEAP, DMF, (16%))→ 304 (81)

305 →(−1.5 V (Ag/AgCl), 10 mol % [Ni(cyclalm)](CLO$_4$), NH$_4$ClO$_4$, 0.1 M TEAP, DMF, (49%))→ 306 (82)

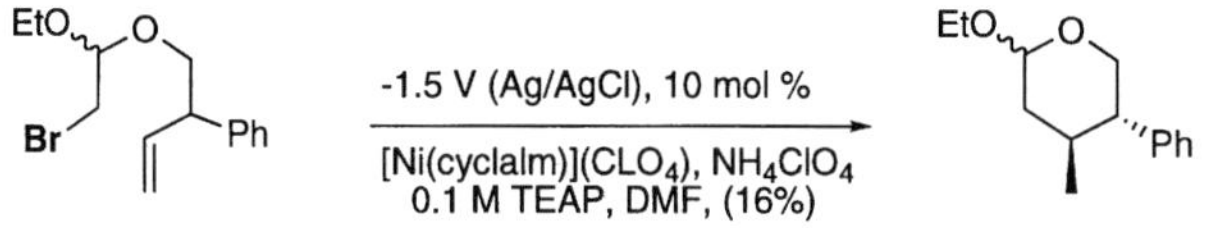

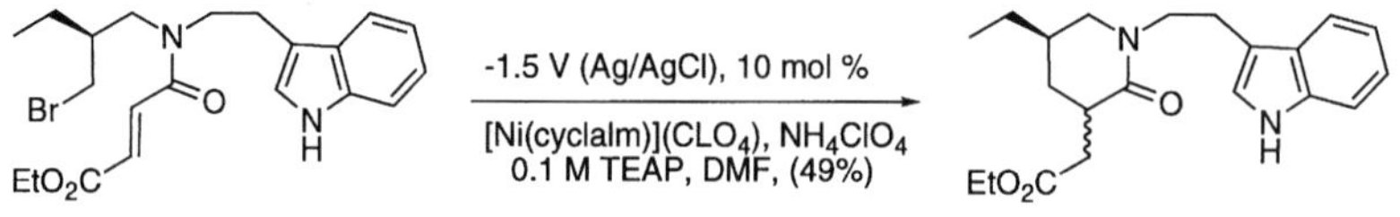

6.5 Cyclization of 1,n-Dihalides

Many reducing agents are capable of severing a carbon-halogen bond. Cathodic cleavage provides perhaps the most versatile method, and has been put to excellent use. The electrochemical variation of the Wurtz reaction constitutes a powerful method for the construction of a variety of rings, particularly strained systems. Dramatic examples are provided by the assembly of bicyclobutane (**308**) [89], bicyclohexene (**310**) [90–92], [2.2.2]propellane (**312**) [93], spiropentane (**316**) [94], β-lactams **318** [95], and a variety of small-ring heterocycles (**320**) [96, 97].

$$\text{Cl-}\underset{307}{\square}\text{-Br} \xrightarrow[-2.0 \text{ V}]{\text{DMF, TEABr}} \underset{308 \ (60\%)}{\square} \tag{83}$$

$$\underset{309}{\overset{\text{Cl}}{\underset{\text{Br}}{\square}}} \xrightarrow[-20 °\text{C, } -2.5 \text{ V}]{\text{TEABr, DMF}} \underset{310}{\square} \xrightarrow{\text{"quantitative"}} \underset{310a}{} \tag{84}$$

$$\underset{311}{\overset{\text{Br}}{\underset{\text{Br}}{}}} \xrightarrow[-15 \text{ to } -25\ °\text{C}]{-2.35 \text{ V, DMF, TEABr,}} \left[\underset{312}{}\right] \longrightarrow \underset{313}{} \tag{85}$$

$$\underset{314}{\overset{\text{BrH}_2\text{C}\quad\text{CH}_2\text{Br}}{\text{BrH}_2\text{C}\quad\text{CH}_2\text{Br}}} \xrightarrow[\text{e (Hg), TBAClO}_4]{-1.8 \text{ V (47\%)}} \underset{315}{\overset{\text{BrH}_2\text{C}}{\text{BrH}_2\text{C}}} \xrightarrow{-2.3 \text{ V (40\%)}} \underset{316}{} \tag{86}$$

$$\underset{317,\ E = \text{CO}_2\text{Et}}{\overset{p\text{-MeOC}_6\text{H}_4}{\underset{\text{O}\quad E\quad\text{CO}_2\text{Et}}{\text{Br}\frown\text{N}\frown\text{Br}}}} \xrightarrow[\text{TEAClO}_4,\ 85\%]{\text{e (Hg), DMF}} \underset{318}{\overset{\text{CO}_2\text{Et}}{\underset{\text{O}}{\text{CO}_2\text{Et}}}\quad\text{C}_6\text{H}_4\text{OMe-}p} \tag{87}$$

$$\underset{\underset{319\text{a-f}}{\text{Br Br}}}{\text{Ph}\diagdown\text{Y}\diagup\text{Ph}} \xrightarrow{2\text{e, } -2\text{Br}^\ominus} \underset{320\text{a-f}}{\text{Ph}\diagdown\text{Y}\diagup\text{Ph}} \quad Y = \text{CH}_2,\ \text{CO, S, SO, SO}_2,\ \text{PO(OEt)}_2 \tag{88}$$

The literature indicates that cyclopropane formation occurs in a stepwise manner [98–100], with the initial generation of a carbanion and subsequent S_N2 displacement. This was ascertained from experiments involving the reduction of *meso*- and *d,l*-2,4-dibromopentane. Each afforded roughly equal amounts of cis- and trans-1,2-dimethylcyclopropane, this indicative of the formation of a carbanion which subsequently undergoes cyclization. Furthermore, reduction of (+)-(2S,4S)-dibromopentane afforded (−)-(1R,2R)-dimethylcyclopropane of high optical purity, indicating clean inversion of configuration at the center being attacked in the cyclization event.

7 Ring Expansion

α-(ω-Bromoalkyl) β-keto esters **321** have also been utilized as substrates for electrochemically promoted reductive cyclizations [101]. Lead appears to be the cathode of choice. The reactions are invariably carried out in the presence of a Lewis acid, including, TMSBr, TiBr(OPr-i)$_3$, and Ti(OEt)$_4$. In their absence, the yields are reduced to values of less than 20%. Unfortunately, their precise role in the chemistry is not clearly defined, though it is suggested that they may assist as a "typical" Lewis acid, in the addition to the carbonyl.

$$321 \xrightarrow[\text{Lewis acid/DMF}]{\text{e (Pb cathode)}} 322 \qquad (89)$$

Three reaction paths are observed, their relative proportions being dependent upon the nature of the substrate. For example, the reduction of **323** affords both the cyclization product **324** as well as **325**, a substance which is clearly formed by reductive cleavage of the carbon-halogen bond without cyclization. The third mode of reaction is exemplified by the nine-membered ring keto ester **326**. Presumably, the initially formed cyclization adduct fragments to afford an ester-stabilized carbanion which subsequently undergoes protonation leading to **327**.

$$323 \xrightarrow[\substack{\text{TMSBr/DMF, 4 F/mol} \\ (65\%\ \mathbf{324},\ 17\%\ \mathbf{325})}]{\text{e (Pb cathode)}} 324 + 325 \qquad (90)$$

$$326 \xrightarrow[\substack{\text{Ti(OEt)}_4/\text{DMF, 3 F/mol} \\ (45\%\ \mathbf{327},\ 18\%\ \mathbf{328})}]{\text{e (Pb cathode)}} 327 + 328 \qquad (91)$$

Analogous transformations have been initiated using alternative methods. Tributyltin hydride and sodium naphthalenide [101], for example, were examined in an effort to probe the possible intermediacy of a radical or carbanion, respectively. The results were compared with those achieved electrochemically. As illustrated, the results were different for each set of reagents, though the sodium naphthalenide and electrochemical results are most similar. This information has been used to suggest that a carbanion is formed electrochemically and participates in the cyclization event.

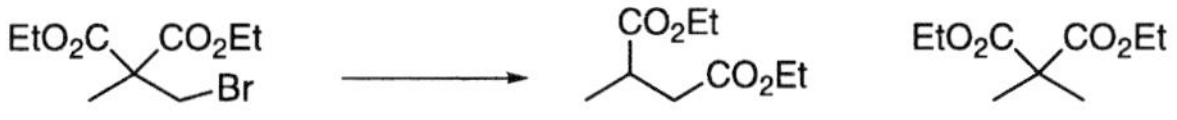

method	% 329	% 330
n-Bu$_3$SnH (radical)	0	100
Na C$_{10}$H$_8$ (carbanion)	100	0
e, Ti(OEt)$_4$	60	5

(92)

8 Reduction of Sulfonamides: Cyclization Promoted by *N*-Tosyl Group Cleavage

The mitomycins (**331**) have attracted much attention as a result of their interesting structure and especially because of their potent anti-cancer properties [102]. One approach to these materials is illustrated in Scheme 22 [103]. A key intermediate in the implementation of this plan is azocinone **333**, a substance

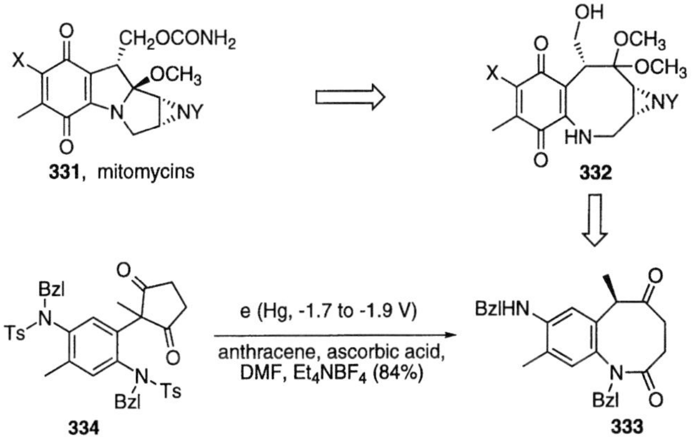

Scheme 22

that proved accessible in high yield via the novel mediated *N*-tosyl group cleavage-cyclization-fragmentation of **334**. Here, ascorbic acid behaves as a proton donor and a reductant in combination with anthracene.

9 Concluding Remarks

Unfortunately, electrochemistry has not been fully embraced by organic chemists, despite the fact that it often provides methodology which may be better suited or superior to a non-electrochemical counterpart. Some of this reluctance is undoubtedly related to the fact that the techniques and equipment are not readily available in each laboratory setting. One hopes that eventually the community will consider electrochemistry simply as another tool, to be explored as routinely as any other. Like all others, it ought to be evaluated in a given context, and used in accordance with its merit.

Cathodic cyclization reactions have supplied and continue to provide a fertile territory for the development and exploration of new reactions and the determination of reaction mechanism. Two areas that appear to merit additional exploration include the application of existing methodology to the synthesis of natural products, and, more significantly, a systematic assessment of the factors associated with the control of both relative and absolute stereochemistry. Until there is a solid foundation to which the non-electrochemist can confidently turn in evaluating the prospects for stereochemical control, it seems somewhat unlikely that electrochemically-based methods will see widespread use in organic synthesis. Fortunately, this comment can be viewed as a challenge and as a problem simply awaiting creative solution.

Acknowledgments. We are grateful to the Electric Power Research Institute (EPRI) and the National Science Foundation (NSF) for a joint program supporting our efforts in organic electrochemistry.

References

1. Gorny R, Schäfer HJ, Fröhlich R (1995) Angew Chem Int Ed Engl 34: 2007
2. Swenton JS, Morrow GW (1991) Tetrahedron 47: 531
3. Shono T (1991) Electroorganic synthesis, 1st edn. Academic Press, London
4. Fry AJ (1989) Synthetic organic electrochemistry, 2nd edn. Wiley, New York
5. Shono T (1984) Electroorganic chemistry as a new tool in organic synthesis, 1st edn. Springer, Berlin Heidelberg New York
6. Baizer MM (1991) In: Lund H, Baizer MM (eds) Organic electrochemistry. Dekker, New York
7. Eberson L, Schäfer H (1971) Organic electrochemistry. In: Boschke F (ed) Topics in current chemistry, vol 21. Springer, Berlin Heidelberg New York, p 5
8. Anderson JD, Baizer MM, Petrovich JP (1966) J. Org. Chem. 31: 3890
9. Petrovich JP, Anderson JD, Baizer MM (1966) J. Org. Chem. 31: 3897

10. Anderson JD, Baizer MM (1966) Tetrahedron Lett.: 511
11. Baizer MM (1963) Tetrahedron Lett.: 973
12. Baizer MM, Anderson JD (1964) J. Electrochem. Soc. 111: 223
13. Baizer MM (1980) Chemtech 10: 161
14. Baizer MM (1964) J. Electrochem. Soc. 111: 215
15. Anderson JD, Petrovich JP, Baizer MM (1969) Electrochemical preparation of cyclic compounds. In: Taylor EC, Wynberg H (eds) Advances in organic chemistry. vol 6. John Wiley & Sons, Inc., New York, p 257
16. House HO (1972) Modern synthetic reactions, 2nd edn. Benjamin, Menlo Park
17. Moëns L, Baizer MM, Little RD (1986) J. Org. Chem. 51: 4497
18. Mandell L, Daley RF, Day RA (1976) J. Org. Chem. 41: 4087
19. Torii S, Okumoto H, Rashid MA, Mohri M (1992) Synlett: 721
20. Shono T, Kise N, Shirakawa E, Matsumoto H, Okazaki E (1991) J. Org. Chem. 56: 3063
21. Little RD, Baizer MM (1989) Enone electrochemistry. In: Patai S, Rappoport Z (eds) The chemistry of enones. John Wiley & Sons Ltd, New York, p 599
22. Fry AJ, Little RD, Leonetti J (1994) J. Org. Chem. 59: 5017
23. Little RD, Fox DP, Van Hijfte LV, Dannecker R, Sowell G, Wolin RL, Moëns L, Baizer MM (1988) J. Org. Chem. 53: 2287
24. Parker VD (1986) Electrochemical applications in organic chemistry. In: Fry AJ, Britton WE (eds) Topics in organic electrochemistry. Plenium, New York, p 35
25. Nadjo L, Savéant JM (1973) J. Electroanal. Chem. 48: 113
26. Parker VD (1980) Acta Chem. Scand. 34B: 359
27. Andrieaux CP, Savéant JM (1974) J. Electroanal. Chem. 53: 165
28. Beckwith ALJ, Ingold KU (1980) In: deMayo P (ed) Rearrangements in ground and excited states, vol 1. Academic Press, New York, Chapter 4
29. Lusztyk J, Maillard B, Deycard S, Lindsay DA, Ingold KU (1987) J. Org. Chem. 52: 3509
30. Tsang R, Dickson JK, Pak H, Walton R, Fraser-Reid B (1987) J. Am. Chem. Soc. 109: 3484
31. Beckwith ALJ, Hay BP (1989) J. Am. Chem. Soc. 111: 230
32. Kochi HF (1984) Acc. Chem. Res. 17: 137
33. Kresge AJ (1975) Acc. Chem. Res. 8: 354
34. Nielson MF, Hammerich O, Parker VD (1986) Acta Chem. Scand. B40: 101
35. Parker VD, Tilset M, Hammerich O (1987) J. Am. Chem. Soc. 109: 7905
36. Bode HE (1989) Ph.D. Dissertation, Ph.D. Dissertation, University of California, Santa Barbara
37. Bode HE, Sowell CG, Little RD (1990) Tetrahedron Lett. 31: 2525
38. Cane DE, Rossi T (1979) Tetrahedron Lett.: 2973
39. Stork G, Maldonado L (1974) J. Am. Chem. Soc. 96: 5272
40. Van Hijfte LV, Little RD, Petersen JL, Moeller KD (1987) J. Org. Chem. 52: 4647
41. Sowell CG, Wolin RL, Little RD (1990) Tetrahedron Lett. 31: 485
42. Kende AS, Roth B, Sanfilippo PJ, Blacklock T (1982) J. Am. Chem. Soc. 104: 5808
43. Little RD, Wolin R, Sowell G (1994) Denki Kagaku 62: 1105
44. Amputch MA, Little RD (1991) Tetrahedron 47: 383. Utley JHP, (1987) Electrogenerated bases. In: Steckhan E (ed) Topics in current chemistry. Springer, Berlin Heidelberg New York, p 133
45. Little RD, Sowell CG (1991) Stereoselection in electroreductive cyclization. Construction of a Corey lactone precurser. In: Little RD, Weinberg NL (eds) Electroorganic synthesis, festschrift for Manuel M. Baizer. Dekker, New York, p 323
46. Shono T, Mitani M (1971) J. Am. Chem. Soc. 93: 5284
47. Shono T, Nishiguchi I, Ohmizu H, Mitani M (1978) J. Am. Chem. Soc. 100: 545
48. Kariv-Miller E, Mahachi TJ (1986) J. Org. Chem. 51: 1041
49. Kariv-Miller E, Nanjundiah C, Eaton J, Swenson KE (1984) J. Electroanal. Chem. 167: 141
50. Loffredo DM, Swartz JE, Kariv-Miller E (1989) J. Org. Chem. 54: 5953
51. Giese B (1986) Radicals in organic synthesis: formation of carbon-carbon bonds, 1st edn. Pergamon Press, Oxford
52. Kashimura S, Ishifune M, Shono T (1995) Electroreduction of organic compounds using chemically reactive Mg electrode. In: Torii S (ed) Novel trends in electroorganic synthesis. Kodansha, Tokyo, p 213
53. Shono T, Masuda H, Murase H, Shimomura M, Kashimura S (1992) J. Org. Chem. 57: 1061
54. Pattenden G, Robertson GM (1983) Tetrahedron Lett. 24: 4617
55. Pattenden G, Robertson GM (1985) Tetrahedron 41: 4001
56. Shono T, Kise N (1990) Tetrahedron Lett. 31: 1303

57. Shono T, Kise N, Fujimoto T, Tominaga N, Morita H (1992) J. Org. Chem. 57: 7175
58. Ohmori H, Maki T, Maeda H (1995) Synthetic application of acyl- and alkoxy-phosphonium ions. In: Torii S (ed) Novel trends in electroorganic synthesis. Kodansha, Tokyo, p 345–348
59. Maeda H, Maki T, Ohmori H (1992) Tetrahedron Lett. 33: 1347
60. Maeda H, Maki T, Ohmori H (1994) Chem. Pharm. Bull. 42: 1041
61. Maeda H, Maki T, Eguchi K, Koide T, Ohmori H (1994) Tetrahedron Lett. 35: 4129
62. Kise N, Suzumoto T, Shono T (1994) J. Org. Chem. 59: 1407
63. Bunce RA (1995) Tetrahedron 51: 13103
64. Kariv-Miller E, Maeda H, Lombardo F (1989) J. Org. Chem. 54: 4022
65. Kariv-Miller E, Lombardo F, Maeda H (1991) The 5,6 vs the 6,5 electroreductive tamdem cyclization of ketones. In: Little RD, Weinberg NL (eds) Electroorganic synthesis, festschrift for Manuel M. Baizer. Marcel Dekker, Inc., New York, p 75
66. Gassman PG, Lee C-J (1994) Synth. Comm. 24: 1465
67. Gassman PG, Lee C (1989) J. Am. Chem. Soc. 111: 739
68. Gassman PG, Rasmy OM, Murdock TO, Saito K (1981) J. Org. Chem. 46: 5455
69. Gassman PG, Lee C (1989) Tetrahedron Lett. 30: 2175
70. Smith AJ, Hannah DJ (1979) Tetrahedron 35: 1183
71. Little RD, Masjedizadeh MR, Wallquist D, McLoughlin JI (1995) The intramolecular Michael reaction In: Paquette LA (ed) Organic reactions. vol 47. John Wiley & Sons, Inc., New York, p 316
72. Nugent ST, Baizer MM, Little RD (1982) Tetrahedron Lett. 23: 1339
73. Amputch MA, Matamoros R, Little RD (1994) Tetrahedron 50: 5591
74. Scheffold R, Dike M, Dike S, Herold T, Walder L (1980) J. Am. Chem. Soc. 102: 3642; see also Nédélec J-Y, Périchon J, Troupel M, Topics Curr. Chem, this issue
75. Scheffold R, Rytz G, Walder L, Orlinski R, Chilmonczyk Z (1983) Pure & Appl. Chem. 55: 1791
76. Scheffold R (1985) Chimia 39: 203
77. Torii S, Inokuchi T, Yukawa T (1985) J. Org. Chem. 50: 5875
78. Inokuchi T, Kawafuchi H, Torii S (1991) J. Org. Chem. 56: 5945
79. Inokuchi T (1995) The electroorganic syntheses by using low valency cobalt complex. In: Torii S (ed) Novel trends in electroorganic synthesis Kodansha, Tokyo, p 223; see also Nédélec J-Y, Périchon J, Troupel M, Topics Curr. Chem, this issue
80. Shao R-l, Cleary JA, La Perriere DM, Peters DG (1983) J. Org. Chem. 48: 3289
81. Shao R-l, Peters DG (1987) J. Org. Chem. 52: 652
82. Willett BC, Moore WM, Salajegheh A, Peters DG (1979) J. Am. Chem. Soc. 101: 1162
83. Moore WM, Salajegheh A, Peters DG (1975) J. Am. Chem. Soc. 97: 4954
84. Mubarak MS, Nguyen DD, Peters DG (1990) J. Org. Chem. 55: 2648
85. Tanaka H, Sumida S, Sorajo K, Torii S (1995) Ni/Pb bimetal-redox mediated reductive addition/cyclization of allenecarboxylate with allyl bromide in a electrolysis media. In: Torii S (ed) Novel trends in electroorganic synthesis. Kodansha, Tokyo, p 193
86. Tanaka H, Kameyama S, Sumida S, Yamada T, Tokumaru Y, Shiroi T, Sasaoka M, Taniguchi M, Torii S (1991) Synlett: 889; Tanaka H, Kameyama Y, Sumida S, Torii S (1992) Tetrahedron Lett. 33: 7029
87. Tanaka H, Ren Q, Torii S, in: Torii S (ed) Novel trends in electroorganic synthesis. Kodansha, Tokyo, p 195
88. Ihara M, Katsumata A, Setsu F, Tokunaga Y, Fukumoto K (1996) J. Org. Chem. 61: 677
89. Rifi MR (1967) J. Am. Chem. Soc. 89: 4442
90. Casanova J, Bragin J, Cottrell FD (1978) J. Am. Chem. Soc. 100: 2264
91. Casanova J, Rogers H (1974) J. Org. Chem. 39: 3803
92. Wiberg KB, Bailey WF, Jason ME (1974) J. Org. Chem. 39: 3803
93. Wiberg KB, Epling GA, Jason M (1974) J. Am. Chem. Soc. 96: 912
94. Rifi MR (1971) J. Org. Chem. 36: 2017
95. Carelli I, Inesi A (1986) Synthesis: 591
96. Fry AJ, Ankner K, Handa V (1981) Tetrahedron Lett. 22: 1791
97. Fry AJ, Chung L-L (1976) Tetrahedron Lett.: 645
98. Wiberg KB, Epling GA (1974) Tetrahedron Lett.: 1119
99. Fry AJ, Britton WE (1971) Tetrahedron Lett. 46: 4363
100. Fry AJ, Britton WE (1973) J. Org. Chem. 38: 4016
101. Shono T, Kise N, Uematsu N, Morimoto S, Okazaki E (1990) J. Org. Chem. 55: 5037
102. Shirahata K, Hirayama N (1983) J. Am. Chem. Soc. 105: 7199
103. Oda K, Ohnuma T, Ban Y (1984) J. Org. Chem. 49: 953

Intramolecular Carbon–Carbon Bond Forming Reactions at the Anode

Kevin D. Moeller

Department of Chemistry, Washington University, St. Louis, MO 63130, USA

Table of Contents

1 Introduction	50
2 The Formation of Heterocycles	52
2.1 Overview	52
2.2 Anodic Amide Oxidations – New Routes to Constrained Peptide Mimetics	53
3 Electrochemically Generated Radicals	55
3.1 Recent Advances Using the Kolbe Electrolysis Reaction	55
3.2 The Oxidation of Malonate Derivatives	57
3.3 Mediated Electrolysis Reactions of Doubly Activated Methylenes	58
4 The Use of Electro-Auxiliaries – Cyclizations Originating from Oxonium Ions to Acyl Anion Equivalents	63
4.1 The Use of Stannyl and Silyl Activating Groups	63
4.2 The Use of Anodically Generated Intermediates for Promoting Cathodic Cyclizations	64
5 Direct Intramolecular Carbon–Carbon Bond Forming Reactions	66
5.1 Coupling Reactions Involving Aryl Rings and Olefins	66
5.2 Intramolecular Anodic Olefin Coupling Reactions	76
6 Conclusions	83
7 References	84

K.D. Moeller

Oxidative cyclization reactions offer a unique opportunity both for generating new carbon–carbon bonds and for gaining insight into the mechanisms that govern a variety of radical, radical cation, and cation intermediates. This review surveys recent developments in the use of electrochemistry as a tool for initiating these intriguing reactions. The transformations examined range from fragmentation reactions resulting in radical cyclization pathways to radical cation-based cycloaddition reactions resulting in the formation of two or more bonds. The products generated range from simple five-, six-, and seven-membered rings to complex bridged and fused polycyclic ring skeletons. Many of the cyclic products obtained either retain the functionality used to intitiate their formation or have a higher level of functionality than that found in the starting material. The degree of functionality in the products would appear to make them ideal candidates for further synthetic development.

1 Introduction

The development of new methods for carbon–carbon (C–C) bond formation lies at the heart of modern synthetic organic chemistry. These transformations are important because they not only allow for the optimization of single steps within a synthetic sequence, but also allow for the exploration of entirely new synthetic strategies for the construction of complex organic molecules. Particularly important are C–C bond-forming reactions that are capable of assembling rings. One technique that has often tempted synthetic organic chemists as a means for effecting these transformations is organic electrochemistry. In principle, electrochemistry can offer a chemist a method for generating C–C bonds from highly reactive intermediates while maintaining reaction conditions that are gentle enough to be compatible with a vast array of functional groups. Add to this the utility of electrochemistry for selectively reversing the polarity of known functional groups at preset potentials and for carrying out oxidation and reduction reactions without the need for stoichiometric chemical reagents and it would seem that electrochemistry would be among the most common synthetic methods studied. Yet in spite of this potential and numerous success stories [1, 2], electrochemistry remains a vastly underutilized tool for synthesis. This is particularly true of anodic oxidation reactions.

It is possible that no electrochemically-based strategy for synthesis has a greater potential for development as a general method than does the anodic initiation of cyclization reactions. Whether effected directly (cf. the anodic coupling of phenols) or indirectly (cf. the generation of α-methoxy amides followed by treatment with acid), these reactions offer the opportunity for generating C–C bonds and constructing rings while either maintaining or increasing the overall functionality of the molecule. In addition, it is tempting to suggest that an electrochemical approach to studying reactions of this type would offer distinct advantages over analogous chemical approaches [3]. Many of the chemical methods for studying oxidative cyclization reactions have been limited by the acidic nature of the oxidizing agent or by the inability of the reagent to oxidize functional groups having a wide range of oxidation potentials. For example, one of the most common reagents for initiating oxidative cyclization reactions is Mn(OAc)$_3$ [3]. Although Mn(OAc)$_3$ provides an excellent method for oxidizing 1,3-dicarbonyl compounds, its use in the oxidation of enol ethers is plagued by competitive hydrolysis of the starting material. Hence, coupling reactions involving an enol ether substrate have often required the use of a second chemical oxidant. Snider and Kwan have demonstrated that the oxidation of silyl enol ethers derived from phenyl ketones with either Cu(II) or Ce(IV) salts can lead to efficient cyclization reactions [4]. However, related cyclization reactions starting with silyl enol ethers derived from dialkyl ketones proceeded in much lower yields. In priniciple, anodic electrochemistry can provide a single method for oxidizing all of these substrates. However, there are still many questions to be answered before electrochemistry can be claimed as a success in this area.

Until about ten years ago, most of the electrochemical work in the area of initiating oxidative cyclization reactions focused upon two types of reactions; the Kolbe electrolysis and the anodic coupling of phenols. Both families of reactions have been extensively reviewed [5,6]. For those readers who are new to the area of electrochemistry, the Kolbe electrolysis normally involves the use of the anode to effect an oxidative decarboxylation reaction. The decarboxylation reaction leads to the formation of a radical which is, in turn, trapped by a second radical generated from the decarboxylation of a second acid. Early on, two types of cyclization reactions were common. Either the two radicals were generated in the same molecule leading to an intramolecular coupling reaction, or one of the radicals underwent an addition reaction to an olefin before undergoing an intermolecular coupling reaction with a second radical generated from a coacid. An example of this latter type of cyclization is illustrated in Scheme 1 [7]. Besides using acetic acid as the second participant in these reactions (R = CH$_3$), Schäfer and coworkers used acids that led to the incorporation of side chains with an ester functional group (R = (CH$_2$)$_4$CO$_2$Me, (CH$_2$)$_2$CO$_2$Me, and CH$_2$CO$_2$Me). These reactions were important because they demonstrated the compatibility of the reactions with ring formation, the versatility of mixed coupling reactions for varying the alkyl group incorporated at the

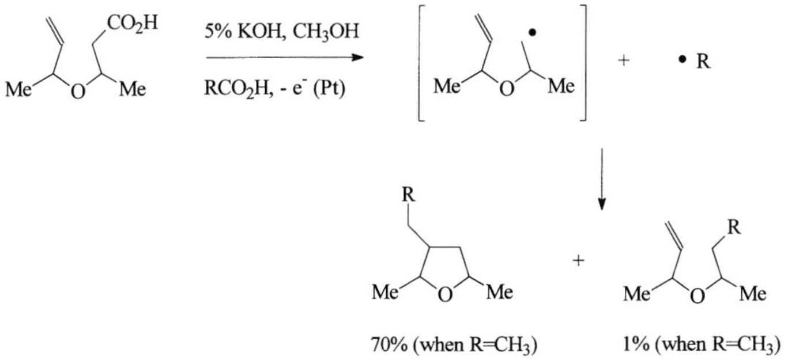

Scheme 1

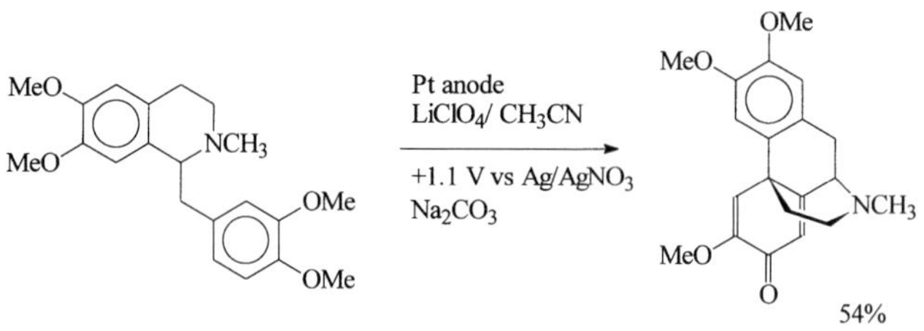

Scheme 2

terminating end of the cyclization, and the utility of electrolysis reactions for generating more than one C–C bond at a time. As we shall see later, the reaction outlined in Scheme 1 paved the way for electrochemically-based tandem cyclization reactions.

Perhaps the most studied anodic cyclization reaction to date has been the coupling of electron-rich aryl rings to form biphenyl-type ring systems. As illustrated in the example outlined in Scheme 2 [8], these reactions proved to be useful tools for constructing polycyclic ring skeletons. In many cases, the yields of these reactions were superior to the yields obtained for alternative chemical procedures. As with the Kolbe electrolysis reaction, these cyclization reactions were important because they not only demonstrated the utility of electrochemistry for generating a specific type of product, but also paved the way for much of the chemistry that followed.

Over the past ten years, work involving the Kolbe electrolysis and the coupling of electron-rich phenyl rings has continued. In addition, several new intramolecular anodic C–C bond-forming reactions have begun to be studied. While the reactions at this point are far from becoming common tools for use by the synthetic organic chemistry community, recent work has clearly begun to demonstrate the role that electrochemistry can play for generating reactive intermediates oxidatively and exploring the ensuing cyclization chemistry. The discussion that follows is aimed at surveying these recent results, as well as highlighting the potential synthetic utility of this growing class of C–C bond-forming reactions.

2 The Formation of Heterocycles

2.1 Overview

From the start, one of the strengths of anodic electrochemistry has been its ability to both generate and modify heterocyclic compounds. Much of this work has involved the use of cyclization reactions and has focused on the

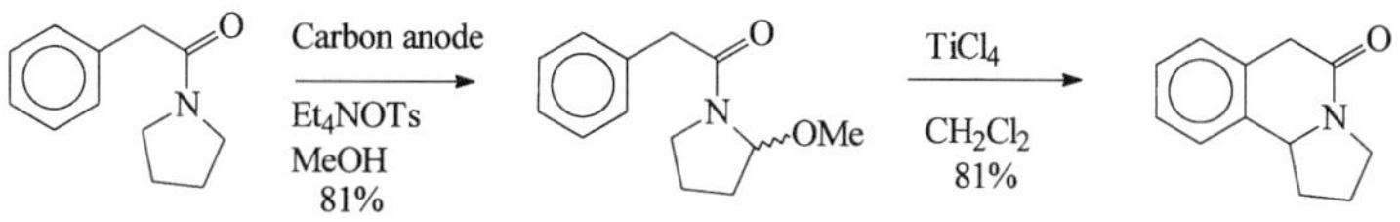

Scheme 3

development of intramolecular routes to the formation of C–X, X–X, and C–C bonds (where X is a heteroatom). While representing a tremendously useful synthetic technique, anodic cyclization reactions leading to the formation of C–X and X–X bonds have been reviewed by Tabakovic [9]. These reactions will not be described again here. Instead, this review will concentrate on anodic routes to heterocyclic rings that involve the formation of new C–C bonds. These approaches span almost the entire range of anodic C–C bond forming reactions, and hence will be discussed below in the context of the reactions used to initiate C–C bond formation.

One approach to the generation of heterocycles that deserves further attention at this point is the anodic oxidation of amides [10]. In 1981, Shono and coworkers demonstrated that a two-step sequence involving the anodic oxidation of an amide could provide a convenient means for preparing polycyclic ring skeletons containing a nitrogen atom (Scheme 3) [11]. While such a route did not involve the direct formation of a C–C bond at the anode, it did demonstrate the utility of anodic electrochemistry for functionalizing molecules and for generating reactive intermediates for use in the construction of C–C bonds. From the start, the anodic amide oxidation reaction was recognized as a synthetic tool with tremendous potential. A number of groups studied similar cyclizations as tools for the total synthesis of natural products. An extensive review of this work has recently appeared [12]. Therefore, a detailed description will not be presented here. However, it is very important to recognize that work aimed at exploiting the utility of these reactions continues in earnest.

2.2 Anodic Amide Oxdations – New Routes to Constrained Peptide Mimetics

Much of the recent work on the use of anodic amide oxidation reactions has focused on the utility of these reactions for functionalizing amino acids and for synthesizing peptide mimetics [13]. For example, in work related to the cyclization strategy outlined in Scheme 3, the anodic amide oxidation reaction has been used to construct a pair of angiotensin-converting enzyme inhibitors [14]. The retrosynthetic analysis for this route is outlined in Scheme 4. In this work, the anodic oxidation reaction was used to functionalize either a proline or a pipercolic acid derivative and then the resulting methoxylated amide used to construct the bicyclic core of the desired inhibitor. A similar approach has recently been utilized to construct 6,5-bicyclic lactam building blocks for

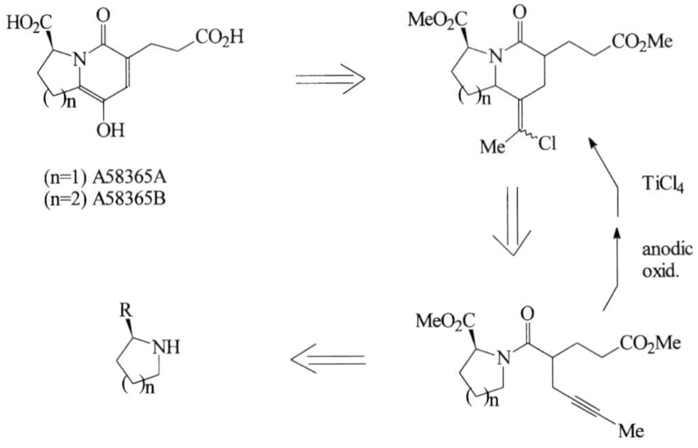

Scheme 4

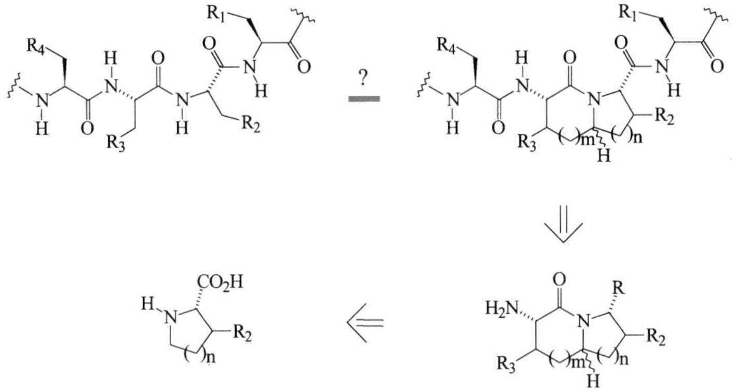

Scheme 5

incorporation into peptides as conformational constraints [15].This route is illustrated in Scheme 5 with a retrosynthetic analysis of the bicyclic lactam-containing peptide fragment. In this example, the proline-based starting material was converted to an amide and then the electrolysis reaction used to functionalize the carbon alpha to the amide nitrogen and to initiate a reaction sequence leading to the construction of the key bicyclic lactam ring skeleton. The effectiveness of electrochemistry for initiating annulation procedures like those outlined in Schemes 4 and 5 has opened up opportunities for designing restricted peptide mimetics by simply "removing" neighboring protons in a desired peptide conformation and then replacing them with a carbon-based bridge. Recent findings have indicated that this approach can lead to hormone analogs with increased affinity and potency [16].

It is important to note that it is the anodic amide oxidation reaction that makes this approach to peptide mimetics possible by allowing for the selective functionalization of amino acid derivatives. At this point, related chemical oxidation reactions have shown neither the generality nor the versatility of the electrochemical reactions for these transformations.

3 Electrochemically Generated Radicals

3.1 Recent Advances Using the Kolbe Electrolysis Reaction

One of the most common methods for electrochemically initiating oxidative cyclization reactions has been to utilize the anode as the source for generating radical intermediates. Like their corresponding chemical relatives [17], electrochemically-generated radicals can be extemely reactive and undergo a variety of olefin addition reactions. As indicated above, most of the early work in this area involved the Kolbe electrolysis. For example, in Scheme 1 the Kolbe electrolysis reaction was used as a method for generating tetrahydrofuran derivatives. Recently, this work has been extended to the synthesis of substituted pyrrolidines (Scheme 6) [18]. In these oxidations, the protecting group of choice for the nitrogen atom was the formyl group. The N-acetoyl group proved to be less deactivating, and in some cases these reactions led to contamination of the product with the N-α-methoxyalkyl amide derived from over-oxidation. As in the earlier furan syntheses, the reactions were compatible with functionality on the second acid which delivers the trapping radical. The reactions were also examined for their compatibility with an acetylene as the terminating group (Scheme 6, Eq. 2). In this example, the reaction did not lead to trapping with the radical derived from the second acid. Apparently the vinyl radical was much more reactive than the methylene radicals and abstracted a hydrogen atom from the solvent.

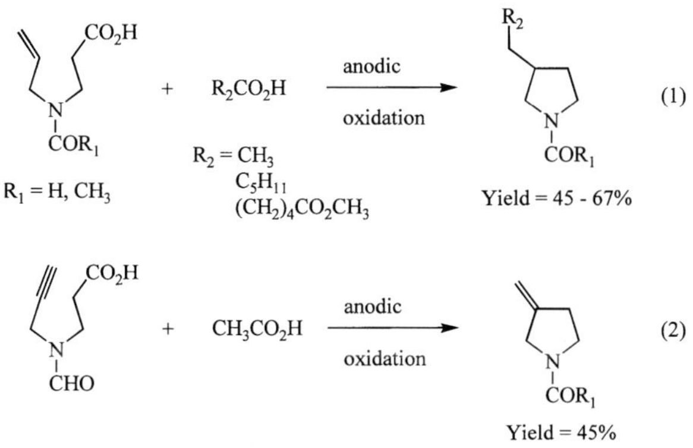

Scheme 6

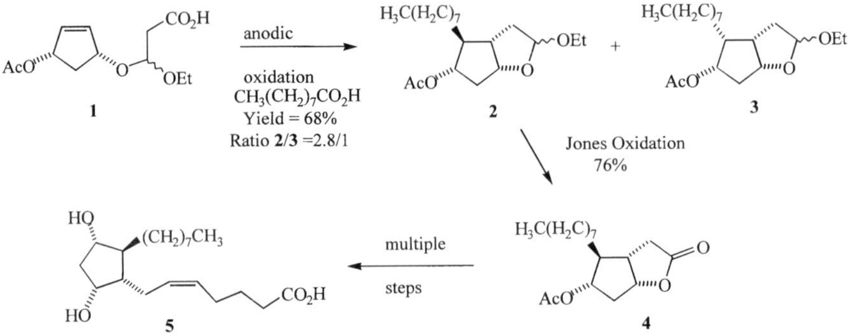

Scheme 7

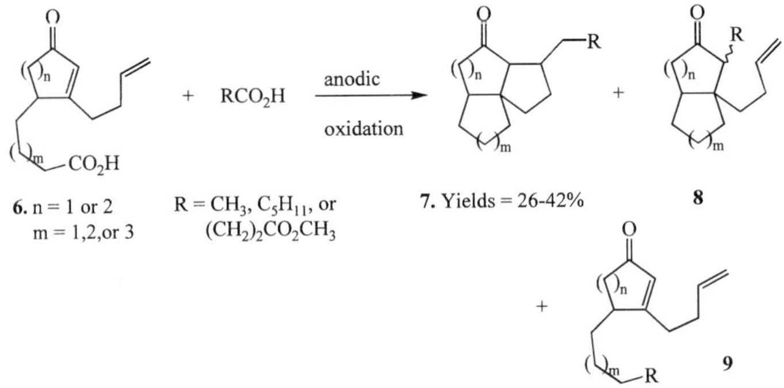

Scheme 8

Kolbe-based cyclization reactions have also been used to construct bicyclic derivatives [18, 19]. For example, Weiguny and Schäfer used the Kolbe reaction to generate a series of advanced prostaglandin precursors. A representative example is illustrated in Scheme 7. A mixture of two bicyclic products was obtained. In all of the cyclization reactions reported, the bicyclic ring possesed *cis*-stereochemistry about the ring juncture, and in all of the cases the major product had the acetoxy group and the newly-added group from the co-acid in a *trans* relationship. In the case illustrated in Scheme 7, the major product was converted into the known prostaglandin precursor **4** by treatment with Jones reagent. Compound **4** had previously been converted into prostaglandin **5** by Kishi and coworkers [20].

The Kolbe electrolysis has also been used to initiate tandem radical cyclization reactions [21]. Recently, Matzeit and Schäfer reported that these reactions could be used to construct angularly fused tricyclic ring skeletons (Scheme 8) [22]. The reaction led to the formation of three new C–C bonds. In addition to the desired tricyclic product, the reaction formed a pair of products having the general structures of **8** and **9**. For example, in the case where n was equal to 2, m was equal to one, and R was a methyl group, the anodic oxidation led to a 42% yield of the tricyclic **7** along with a 15% yield of **8**, and an 8% yield of **9**. Product **8** was formed by competitive trapping of the intermediate radical species generated from the initial cyclization reaction by the methyl radical derived from oxidation of the co-acid. Product **9** was derived from a similar trapping reaction of the initially formed radical. The best yields of tricyclic product were obtained when the current density was kept at a low level (25 mA/cm^2). In this way, a low concentration of radicals could be maintained at the anode surface. Higher current densities led to significant decreases in the yield of the desired tricyclic product along with an increase in the formation of **8** and **9**. Although the yields of these reactions were moderate, they served to demonstrate the potential for anodic electrochemistry to be used for the generation of multiple C–C bonds in a sequential fashion.

3.2 The Oxidation of Malonate Derivatives

In recent years, the oxidation of 1,3-dicarbonyl compounds has proven to be an effective means for generating radical intermediates and initiating cyclization reactions. As mentioned above, chemical variants on this process have been explored in detail [3]. These reactions often utilized Mn(OAc)$_3$ as the oxidant (Scheme 9).

Similar oxidative cyclization reactions involving the direct oxidation of acyclic 1,3-dicarbonyl compounds have not been reported. However, the generation of radical intermediates by the direct oxidation of cyclic 1,3-dicarbonyl compounds at an anode surface has been reported. Yoshida and coworkers have shown that the anodic oxidation of cyclic 1,3-dicarbonyl compounds in the presence of olefin trapping groups gives rise to a net cycloaddition reaction (Scheme 10) [23]. These cycloaddition reactions proceeded by initial oxidation of the 1,3-dicarbonyl compound at the anode followed by a radical addition to the second olefin. Following a second oxidation reaction, the material then

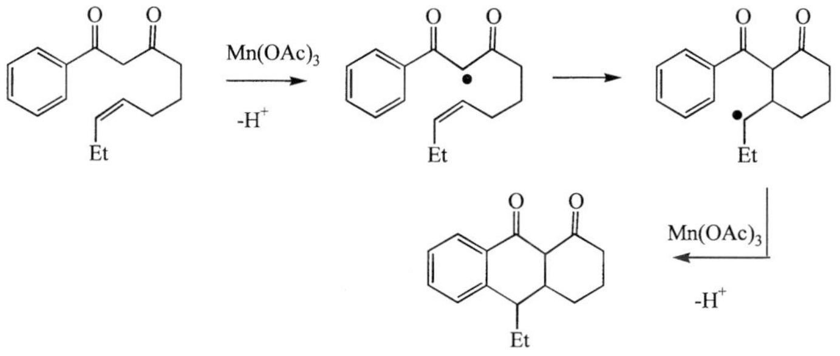

Scheme 9

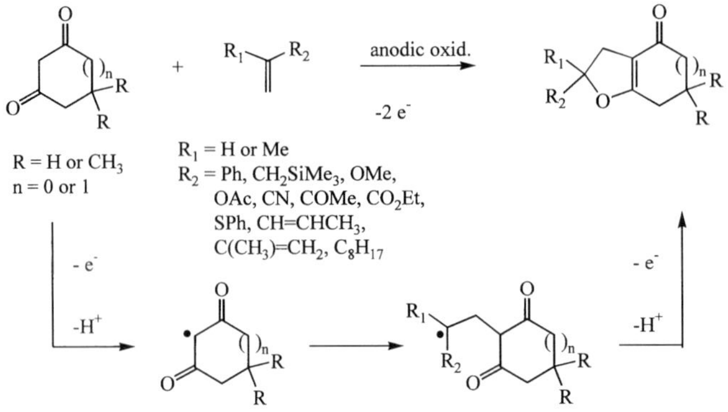

Scheme 10

cyclized to furnish the cycloaddition product. Note that both a new C–C bond and a new C–O bond were formed during this process. The reactions proved to be compatible with a variety of substituents on the olefin. Styrene derivatives, enol ethers, enol esters, allyltrimethylsilane, and 1,3-dienes were all found to be good participants in the reaction. For most of the cyclic 1,3-diketones studied, the highest yield of cycloadduct was obtained when styrene was used as the olefin. The use of simple alkyl substituted olefins and electron-deficient olefins led to much poorer yields of cycloadduct. A number of the cycloaddition reactions also proved compatible with the use of a 2-substituted, cyclic 1,3-diketone. These reactions led to the product even though the formation of a new C–C bond required the formation of a quaternary carbon. Finally, Yoshida and coworkers found that the reactions were not compatible with the use of an acyclic 1,3-diketone. In this example, a complex mixture of products was obtained without the formation of any cycloadduct.

3.3 Mediated Electrolysis Reactions of Doubly Activated Methylenes

Although cyclizations from the direct anodic oxidation of acyclic 1,3-dicarbonyl compounds have not been reported, the analogous mediated reactions have been studied [24]. Snider and McCarthy compared oxidative cyclization reactions using a stoichiometric amount of $Mn(OAc)_3$ with oxidations using a catalytic amount of $Mn(OAc)_3$ that was recycled at an anode surface (Scheme 11). In the best case, the anodic oxidation procedure led to a 59% yield of the desired bridged bicyclic product with the use of only 0.2 equivalents (10% of the theoretical amount needed) of $Mn(OAc)_3$. Evidence that the reaction was initiated by the presence of the mediator was obtained by examining the electrolysis reaction without the added $Mn(OAc)_3$. In this case, none of the cyclized product was obtained. For comparison, the oxidation using

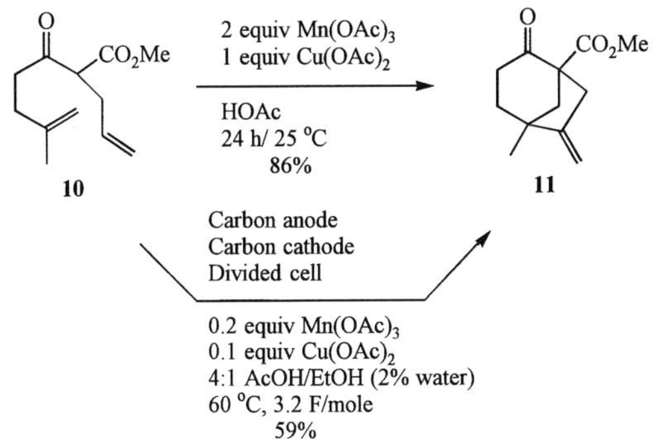

Scheme 11

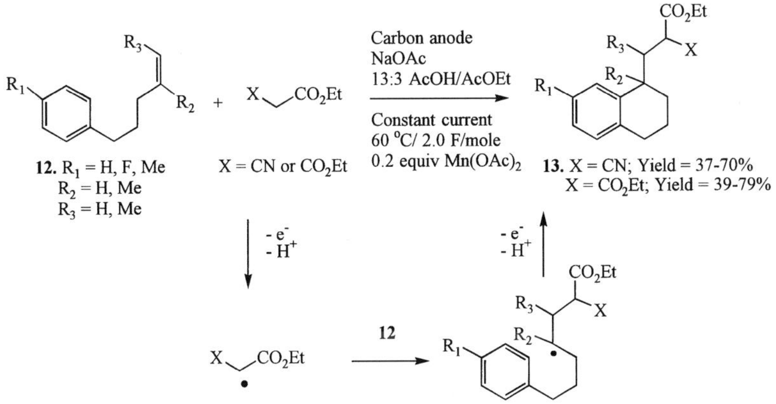

Scheme 12

stoichiometric Mn(OAc)$_3$ led to the formation of an 86% yield of product. It should be noted that the Cu(OAc)$_2$ was added to these reactions in order to assist in the oxidation of the radical intermediates generated from the cyclyzation reactions. The Cu(OAc)$_2$ was then regenerated by the Mn(OAc)$_3$. In general, the reactions using a stoichiometric amount of Mn(OAc)$_3$ were cleaner and led to superior yields of the desired product. In addition to the electrochemical method for regenerating the Mn(OAc)$_3$, chemical oxidants were examined for their effectiveness in recycling the Mn(OAc)$_3$. The best of these reactions used NaIO$_4$ as the stoichiometric oxidant. This reaction was run in DMSO solvent at 60 °C, utilized 0.1 equivalents of Mn(OAc)$_3$ and 0.2 equivalents of Cu(OAc)$_2$, and led to yields of cyclized product that were comparable to the electrochemical oxidation (60% for the oxidation of **10**).

In a related set of experiments, Nishiguchi and coworkers studied oxidative cyclization reactions that were initiated by the addition of radicals derived from the oxidation of an activated methylene group to an olefin (Scheme 12) [25].

These reactions again used Mn(OAc)$_3$ as a mediator for the anodic oxidation reaction. In this case, the Mn^{3+} species was generated by the in situ oxidation of Mn(OAc)$_2\cdot$4H$_2$O. Both diethylmalonate and ethyl cyanoacetate were used as the activated methylene compound. In the case of ethyl cyanoacetate, the yield of product ranged from 53–70%, except for a case where R$_3$ in **12** was equal to a methyl group (37%). A similar observation was made for the reactions involving diethylmalonate. In these cases, the yield of product ranged from 60–73%, except for the case where R$_3$ was equal to methyl (39%). Apparently, product formation was hindered by the increased steric congestion about the forming bond during the addition step.

Nishiguchi and coworkers were also able to use the Mn(OAc)$_3$-mediated oxidation of 1,3-dicarbonyl derivatives as a method to effect furan formation (Scheme 13) [26]. The use of the mediator allowed for the earlier chemistry explored by Yoshida (Scheme 10) to be extended to the use of acyclic 1,3-dicarbonyl derivatives.

In a related experiment, a radical intermediate derived from the oxidation of an activated methylene group was used to trigger a transannular cyclization across an eight-membered ring (Scheme 14) [26]. After cyclization, the reaction

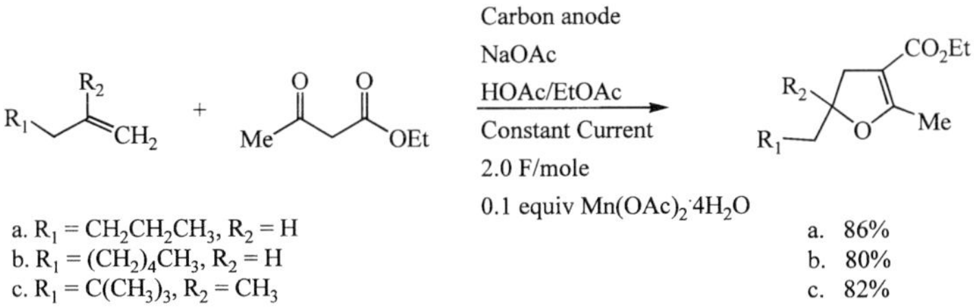

a. R$_1$ = CH$_2$CH$_2$CH$_3$, R$_2$ = H   a. 86%
b. R$_1$ = (CH$_2$)$_4$CH$_3$, R$_2$ = H   b. 80%
c. R$_1$ = C(CH$_3$)$_3$, R$_2$ = CH$_3$ c. 82%

Scheme 13

X = CN or CO$_2$Et

X = CN; Yield = 76%
X = CO$_2$Et; Yield = 78%

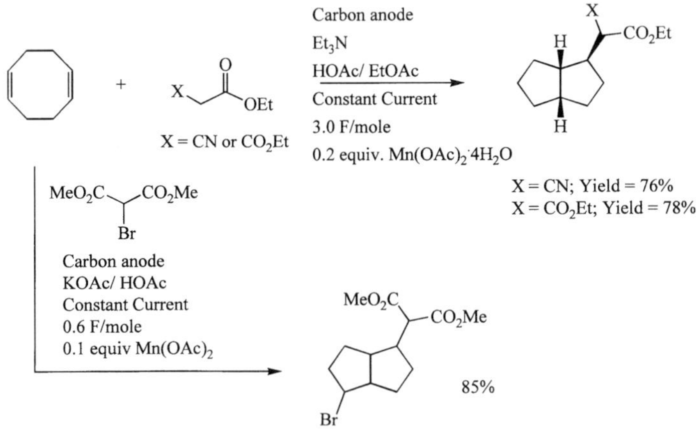

Scheme 14

was terminated by the abstraction of a hydrogen atom. Interestingly, the bicyclo[3.3.0]octane product was obtained as a single diastereomer. A similar sequential addition-transannular cyclization reaction has been studied by Nédélec and Nohair [27]. In this case, the starting material for the oxidation step was dimethyl bromomalonate. The use of the brominated starting material led to bromination at the terminating end of the cyclization reaction. The transannular cyclization again led to *cis*-stereochemistry about the newly formed ring fusion. The stereochemistry of the bromide and malonate substituents was not specified.

Bergamini and coworkers investigated a series of related addition-cyclization reactions [28]. These reactions originated from the oxidation of diethyl 2-phenylmethylmalonate **14** (Scheme 15). The corresponding radical intermediate then underwent addition reactions with both olefin and acetylenic trapping groups. The dependence of the reaction on the nature of the redox couple used as the mediator was examined. Mn^{III}, Ce^{IV}, Pb^{IV}, Co^{III}, and Ni^{III} salts were examined. In all cases, the reactions led to the formation of a bicyclic product and the generation of two new C–C bonds in a fashion directly analogous to the formation of product **13** above (Scheme 12). Reactions using an acetylenic trapping group paralleled the reactions using olefin trapping groups. The use of an acetylene led to products with a higher degree of unsaturation. Of the redox couples studied, Mn, Co, and Ce couples were found to be the most efficient. The presence of a mediator was found to be essential for the formation of product. Attempts at accomplishing the desired transformation using a direct oxidation at the anode surface led to none of the bicyclic product. Finally, the electrochemically based oxidation reactions were compared with the corresponding stoichiometric chemical reactions. In general, the two types of reactions were found to be similar.

The mediated oxidations of α-nitro ketones and α-nitro amides in the presence of olefins have been studied [29]. In this work, Warsinsky and Steckhan found that the ensuing addition-cyclization sequence led to the

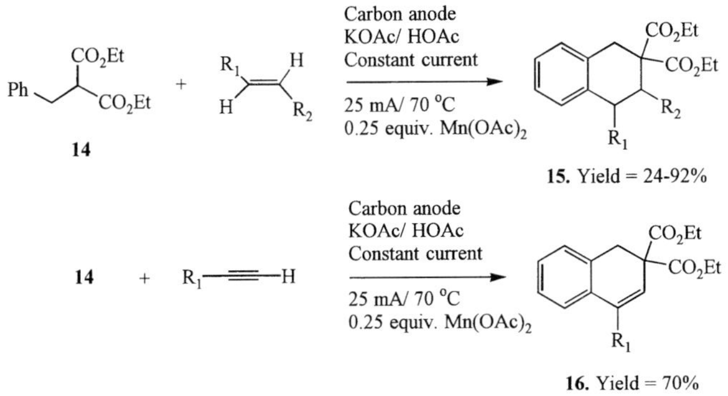

Scheme 15

formation of an isoxazoline N-oxide product (Scheme 16). The electrochemically based oxidation reactions were compared with the corresponding stoichiometric chemical oxidation reactions. In this case, the electrochemical approach led to similar or in some cases slightly better yields than the reactions using a stoichiometric amount of the chemical oxidant. As in earlier reactions, none of the cycloaddition product was obtained from these reactions in the absence of the mediator.

Finally, Nikishin and coworkers have reported that the mediated oxidations of doubly activated methylene compounds can be used to synthesize cyclopropane derivatives (Scheme 17) [30]. Reactions using dimethyl malonate, ethyl cyanoacetate, and malononitrile were studied. Metal halides were used as mediators. When the activated methylene compound was oxidized in the absence of a carbonyl compound, three of the substrate molecules were coupled together to form the hexasubstituted product. Interestingly, when the ethyl cyanoacetate substrate was used the product was formed in a stereoselective fashion (**18b**). In an analogous reaction, oxidation of the activated methylene compounds in the presence of ketones and aldehydes led to the formation of cyclopropane products that had incorporated the ketone or aldehyde (**20**). In the case of **19a**, the reactions typically led to a mixture of stereoisomers.

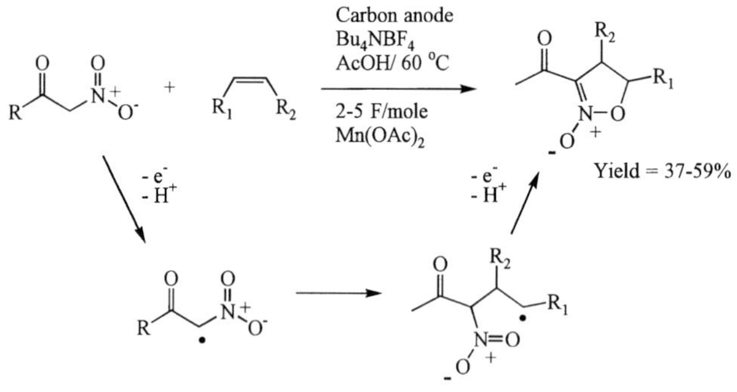

Scheme 16

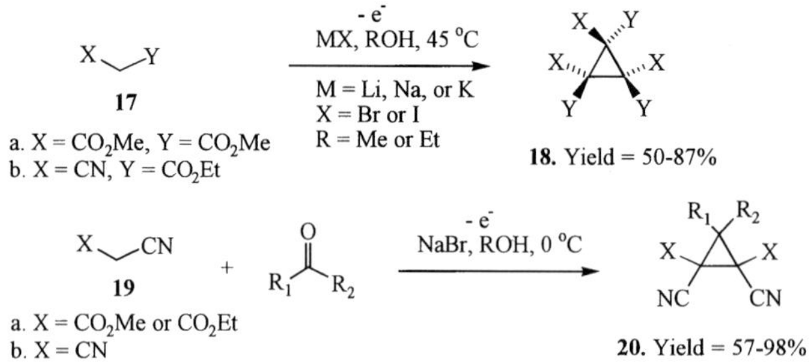

Scheme 17

4 The Use of Electro-Auxiliaries – Cyclizations Originating from Oxonium Ions to Acyl Anion Equivalents

4.1 The Use of Stannyl and Silyl Activating Groups

Recently, Yoshida and coworkers reported that the anodic oxidation of group 14 organometallic compounds can provide a novel route to the formation of new C–C bonds and the construction of new ring systems [31, 32]. In their initial studies, the oxidation of an α-stannyl ether was used to generate an oxonium ion. The presence of the stannyl group enabled the oxidation reaction by substantially lowering the oxidation potential of the oxygen moiety [33]. The oxonium ion was then trapped by an olefin to form the C–C bond and give rise to a cyclic ether product (Scheme 18). In all cases, the cyclization reaction led to an endocyclic product indicating that it was indeed a cationic cyclization. A competitive radical cyclization would have been expected to lead to five-membered ring formation in the cases where n was equal to one. The reaction was terminated by the trapping of the cyclized product with a fluoride ion that was derived from the electrolyte used. Both tetrabutylammonium tetrafluoroborate and tetrabutylammonium hexafluorophosphate electrolytes served as good sources of fluoride.

While most of the examples studied employed a stannyl group as the group 14 metal, the use of an α-silyl ether was also examined. In this case, the same cyclic product was formed as in the corresponding α-stannyl ether substrate. However, the yield of product obtained was higher when the tin based starting material was employed (83% vs 66%).

The reactions were also screened for their ability to generate a five-membered ring. However, this example led to a complex mixture of products.

The reactions did prove to be compatible with the use of α-stannyl amides. In these cases, the cyclization reaction involved the trapping of an N-acyliminium ion by the olefin and led to the formation of substituted piperidines.

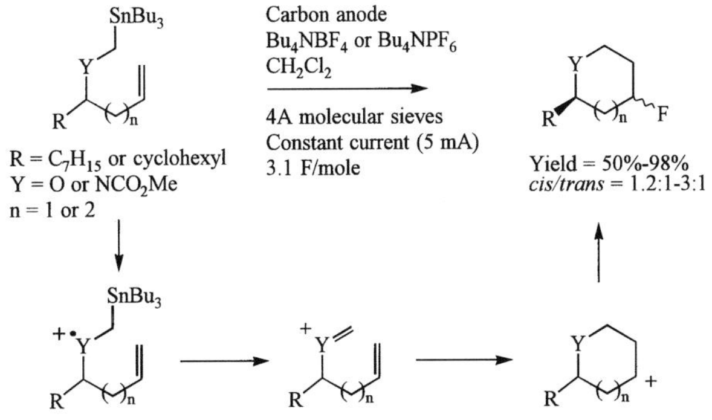

R = C_7H_{15} or cyclohexyl
Y = O or NCO_2Me
n = 1 or 2

Carbon anode
Bu_4NBF_4 or Bu_4NPF_6
CH_2Cl_2
4A molecular sieves
Constant current (5 mA)
3.1 F/mole

Yield = 50%–98%
cis/trans = 1.2:1–3:1

Scheme 18

While termination of the cyclization reactions with fluoride was an interesting and potentially important result, the authors were also interested in trapping the cyclized carbocation with a more synthetically versatile group. One logical choice was to trap with bromide. In principle, the bromide could be incorporated by simply changing the electrolyte to Bu_4NBr. However, this approach failed because the reaction led to selective oxidation of the bromide. To solve this problem, the authors developed a method to generate only low concentrations of bromide during the reaction. This was accomplished by utilizing dibromomethane as the solvent for the reaction. The bromide ion was then generated by reduction of the solvent at the cathode. This technique led to good yields of the desired cyclized product. An example of this transformation is illustrated in Scheme 19 (Eq. 1) [32]

Finally, the reactions were examined in order to determine their compatibility with the initiation of Friedel-Crafts cyclizations (Scheme 19, Eq. 2). Both the use of an α-stannyl ether and an α-stannyl amide substrates led to cyclized product.

The success of these reactions was intriguing because, under normal reaction conditions, group 14 organometallic compounds serve as carbanion synthons. The anodic oxidation reaction successfully reversed this polarity thereby expanding the overall synthetic utility of group 14 organometallic intermediates.

4.2 The Use of Anodically Generated Intermediates for Promoting Cathodic Cyclizations

Recently, Ohmori and coworkers have used an anodic oxidation reaction to promote the reduction of an acid [34]. In this experiment, the anodic oxidation of triphenyl- or tributylphosphine in the presence of a carboxylic acid led to the formation of an acyl phosphonium ion. The acyl phosphonium ion was then reduced at the cathode to form an ylide which then trapped a second carbonyl

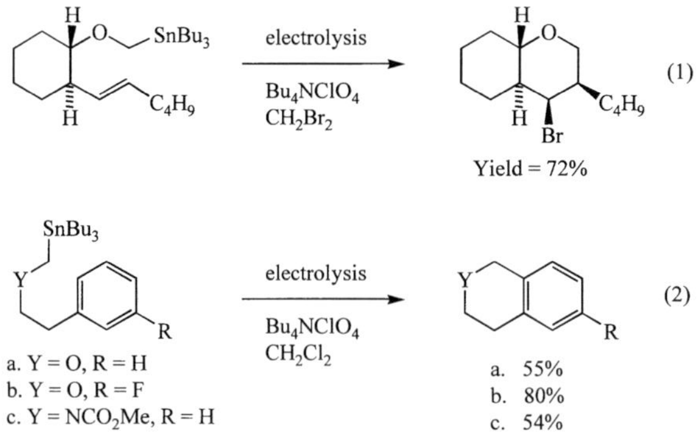

Scheme 19

group in order to generate a new C–C bond and give rise to the cyclization reaction (Scheme 20).

In a related study, the oxidation–reduction sequence was carried out in the presence of an olefin (Scheme 21). Two products were formed. The major product resulted from the net reduction of the carboxylic acid to an aldehyde. The minor product resulted from trapping of the radical anion intermediate generated from the reduction reaction by the olefin. It should be noted that, in the absence of a trapping group, the acid can be selectively reduced to the aldehyde without any over-reduction. Although not in the scope of this review, this is a very useful transformation in its own right [35]. At this time, the yields of the cyclized products from the cyclization reaction of the radical anion with the olefin remain low.

Although not yet optimized for C–C bond formation, these reactions demonstrated the potential for using mixed anodic oxidation - cathodic reduction sequences for initiating transformations that would be difficult to accomplish using more conventional chemical routes. It is hard to imagine putting a strong

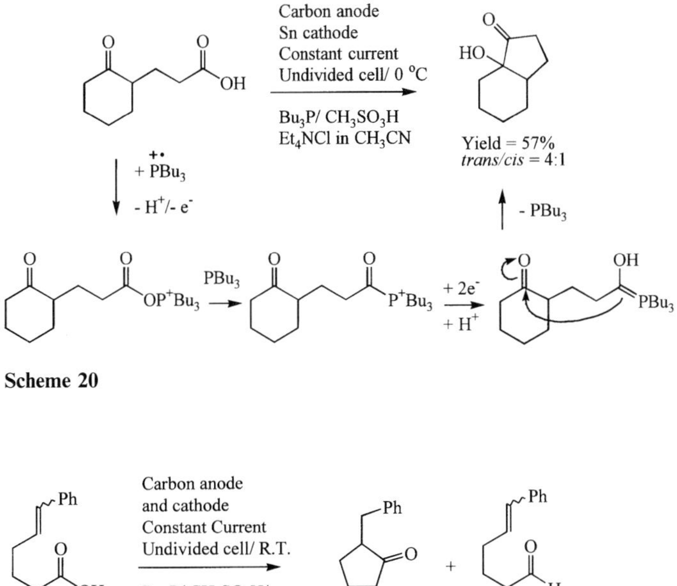

Scheme 20

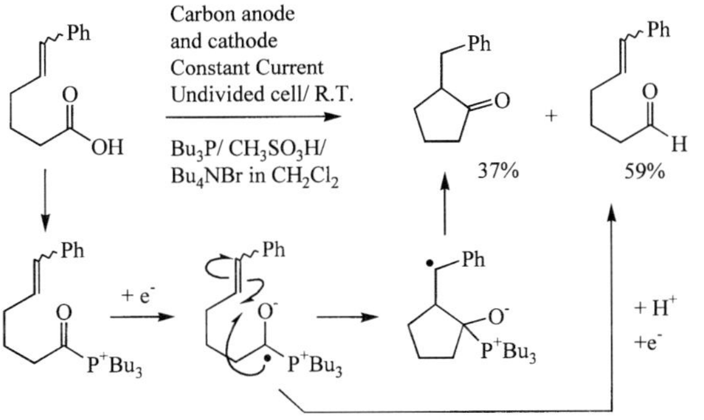

Scheme 21

chemical oxidant and a strong chemical reductant together in the same reaction vessel and winding up with something useful. Clearly, electrochemistry is the ideal method for initiating tandem oxidation–reduction sequences in a one-pot reaction.

5 Direct Intramolecular Carbon–Carbon Bond Forming Reactions

As mentioned in the introduction, the intramolecular coupling of electron-rich phenyl rings has been one of the most successful anodic cyclization reactions studied to date. These reactions have been extensively reviewed [8] and will not be detailed here. However, it is important to note that the general utility of intramolecular coupling reactions of aryl rings did much to illustrate the synthetic versatility of directly forming C–C bonds at anode surfaces. Over the last ten years, much of the work aimed at exploring anodic cyclization reactions has focused on exploiting and expanding this general approach.

5.1 Coupling Reactions Involving Aryl Rings and Olefins

One of the most synthetically useful anodic C–C bond forming reactions developed to date involves the intramolecular coupling of phenol derivatives with olefins. Yamamura has demonstrated that these reactions tend to lead to three classes of products (Scheme 22) [36]. The type of product generated depended strongly on both the nature and stereochemistry of the functional groups attached to the olefin moiety. For example, consider the two cyclization reactions illustrated in Scheme 23 [37]. In this experiment, the stereochemistry of the initial olefin substrate completely dictated the ring skeleton of the product.

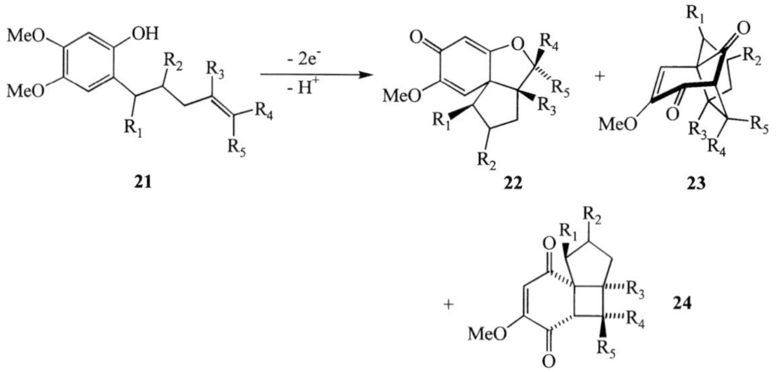

Scheme 22

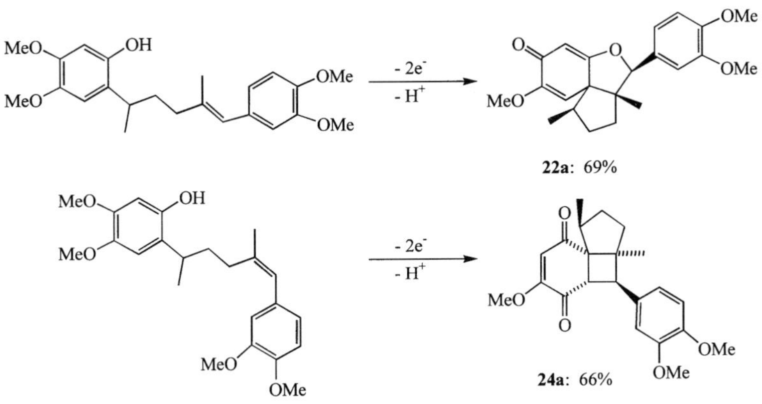

Scheme 23

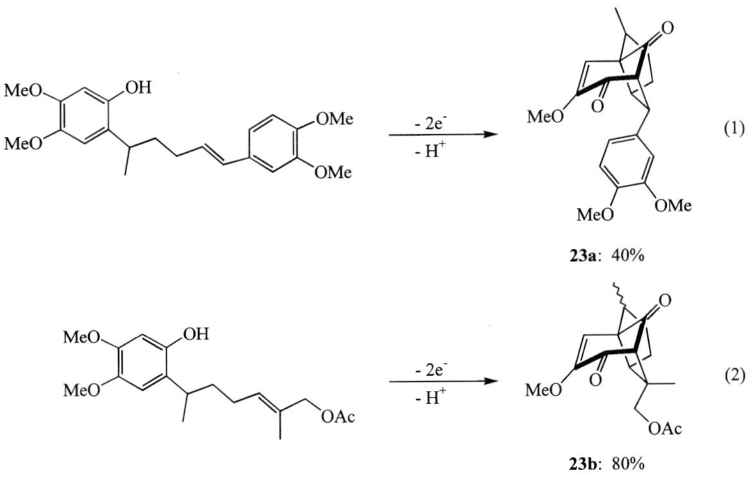

Scheme 24

An E double bond led to a product of type **22** and a Z double bond led to a product of type **24**. The reactions were not just sensitive to stereochemistry. For example, removal of the methyl substituent on the olefin caused the oxidative coupling reaction to lead to a product having the overall structure type **23** (Scheme 24) [38, 39]. A variety of substituted aryl rings were studied (range of yields = 40–63%). As long as the methyl group was not present on the internal carbon of the double bond, the reaction led to the bridged bicyclic ring skeleton. This also proved to be true for cyclization reactions that did not have an aryl ring on the olefin (Scheme 24, Eq. 2) [39, 40].

If the olefin was electron-poor, then the reactions could be channeled to products of type **24** even without the methyl group on the internal carbon of the olefin (Scheme 25) [41].

Both electrolysis products of type **23** and type **24** have proven to be versatile synthetic intermediates. The bridged bicyclic products (**23**) have been used to make natural products containing bicyclo[3.2.1]octene ring skeletons (Scheme 26, Eq. 1) [40] angularly fused tricyclopentanoid ring skeletons (Scheme 26, Eq. 2) [38, 42] and tricyclo[5.3.1.0^{1,5}]undecene ring skeletons (Scheme 26, Eq. 3) [38, 39]. Several "typical" retrosynthetic analyses are outlined below. In all three examples, the brief retrosynthetic analysis presented was designed to emphasize the structure of the target molecule and its relationship to the product and starting material of the electrolysis reaction. In the final two equations, the synthesis depends on the selective fragmentation of the bicyclic ring skeleton. The fragmentation reactions were controlled by manipulating the functionality of the initially formed electrolysis product. As discussed earlier, one of the advantages of constructing molecules by oxidative cyclization reactions is that the reactions either preserve or raise the level of functionality in the

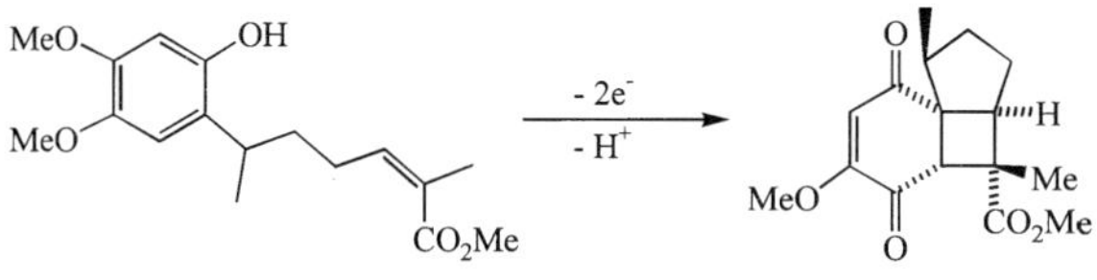

Scheme 25

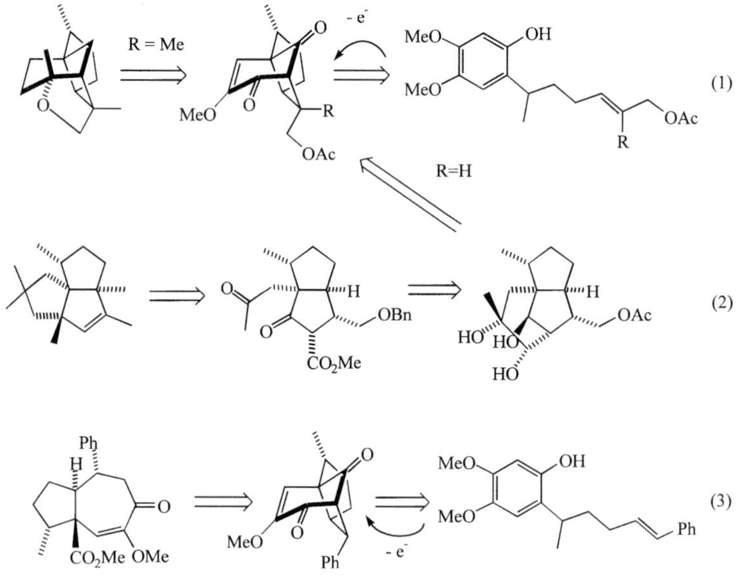

Scheme 26

molecule. As illustrated here, this characteristic of the cyclizations can play a crucial role in the subsequent elaboration of the product.

In a similar fashion, electrolysis products like **24** have been converted to both angularly and linearly fused tricyclopentanoid ring systems [37, 43]. A retrosynthetic analysis for both ring systems is outlined in Scheme 27. The two synthetic routes have the product of the anodic coupling reaction as a common intermediate. Both routes rely on a ring expansion of the four-membered ring to form the central five-membered ring of the product. In each case, a different bond was migrated. The nature of the ring system formed depended on the substitution pattern of the substrate for the migration reaction. Again, the success of the syntheses relied on the ability of the anodic coupling reactions to generate highly functionalized products.

In an effort to explore the factors that govern anodic C–C bond formation, Swenton and coworkers have also been exploring the intramolecular coupling of phenols and olefins (Scheme 28) [44]. In these reactions, initial oxidation of the phenol followed by loss of a proton and a second oxidation led to the formation of a cationic intermediate (**26**). This intermediate was trapped by the olefin to form a second cation that was in turn trapped by methanol to form the final product **28**. When R_2 was equal to methyl (**25b**) or phenyl (**25c**) the reaction led to a good yield of the cyclized product. Reactions where the R_2 was equal to a hydrogen (**25a** and **25d**) were not so successful. The cyclizations were compatible with the incorporation of the olefin into a third ring (**25e**).

Substrate **25b** was used in order to study the effect of various reaction conditions on the yield of the cyclization reaction. The reactions were found to depend on the solvent system used, the current density (the best results were obtained with a current density of 0.84 mA/cm^2), the anode material, and the presence of a mild acid.

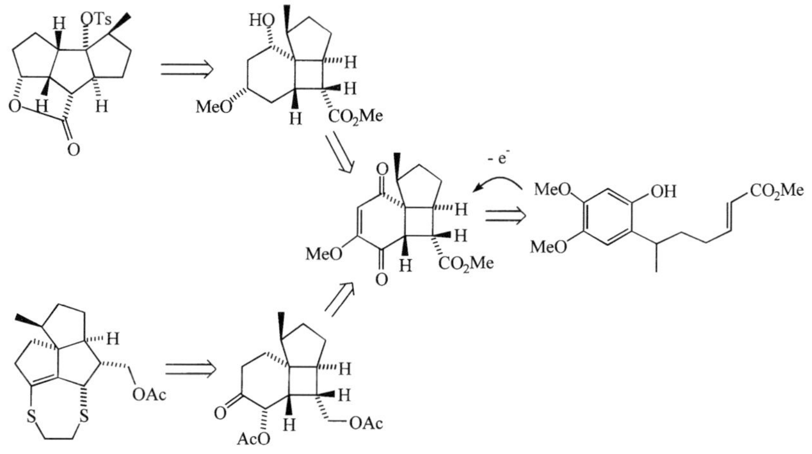

Scheme 27

One of the questions asked about these reactions focused on why the methyl group on the internal carbon of the olefin (R_2) was so important for the cyclization [45]. There appeared to be two possible explanations. Either the methyl group increased the nucleophilicity of the olefin by stabilizing the formation of intermediate carbocation **27**, or the presence of the methyl group led to a conformational change that increased the rate of the cyclization reaction. This second possibility was very intriguing. One could imagine that in the absence of the methyl group the olefin would prefer an orientation that was away from the 4-position of the starting phenol ring and hence distal from the cationic center in intermediate **26**. In short, in the absence of the methyl group conformation **26'** would be favored (Scheme 29). Since the transition state for the cyclization would have to resemble **26''**, the reaction would have an additional steric barrier associated with bringing the olefin close to the carbocation intermediate. When R_2 was equal to a methyl group, the energetic preference for **26'** would be removed. The reaction would no longer have to "pay a price" for bringing the olefin close to the carbocation in the transition state leading to cyclization. In order to see if this second alternative was a possibility, Swenton

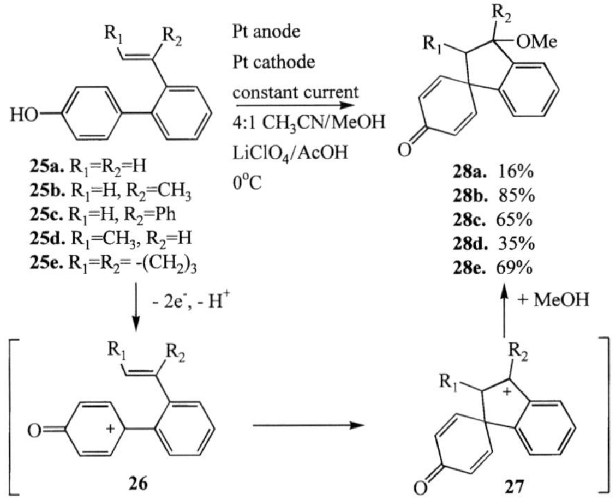

Scheme 28

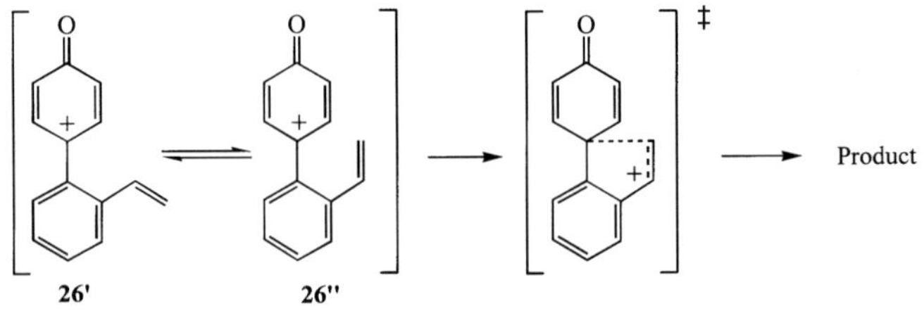

Scheme 29

and coworkers designed a substrate that would "buttress" the olefin and remove the conformational preference inherent with an olefin having R_2 equal to a hydrogen (Scheme 30) [45]. In this example, a methyl group was placed on the phenyl ring *ortho* to the vinyl substituent. This methyl group had the net effect of removing the steric preference favoring a conformation like **26′**. The anodic oxidation of this substrate led to a good yield of cyclized product. Clearly, a substituent on the olefin was not needed for the cyclization reaction in an appropriate system.

A closely related reaction having the phenol protected with a trimethylsilyl group was also examined (Scheme 31) [45]. Unlike the earlier examples, the cyclization reaction resulting from this substrate did not require the presence of a mild acid. The anodic oxidation in methanol solvent with no acetic acid led to a 73% yield of the tricyclic product. In a nearly identical reaction, an anodic oxidation of the trimethylsilyl-protected substrate in the presence of 2,6-lutidine led to the cyclized product in a 60% yield. The use of the silyl group expanded the utility of the anodic C–C bond-forming reaction being studied by allowing for the use of neutral and basic conditions. Hence, it would appear that the cyclization reactions are compatible with the presence of both base and acid sensitive functionality.

In addition to these studies, Swenton and coworkers have recently explored the effects of alkoxy substituents on the reactions [46]. The presence of a single methoxy group clearly enhanced the cyclization reaction (Scheme 32). For example, consider the oxidation of substrate **29a**. The presence of a methoxy group *para* to the vinyl substituent raised the yield of cyclized product from 16% (**25a** above) to 47%. The methoxy group *para* to the vinyl substituent also proved to be compatible with the oxidative cyclization reaction having a methyl

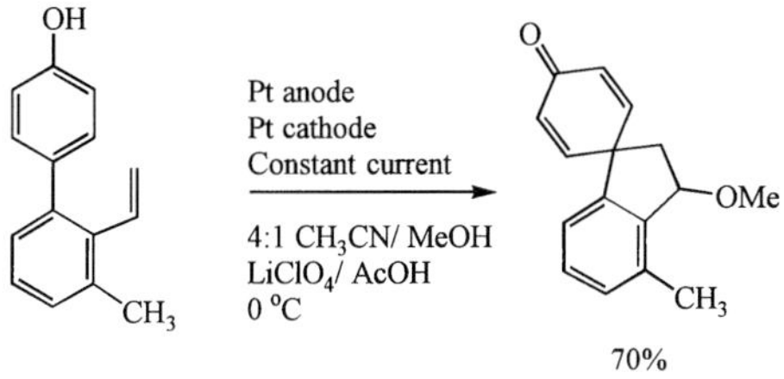

Scheme 30

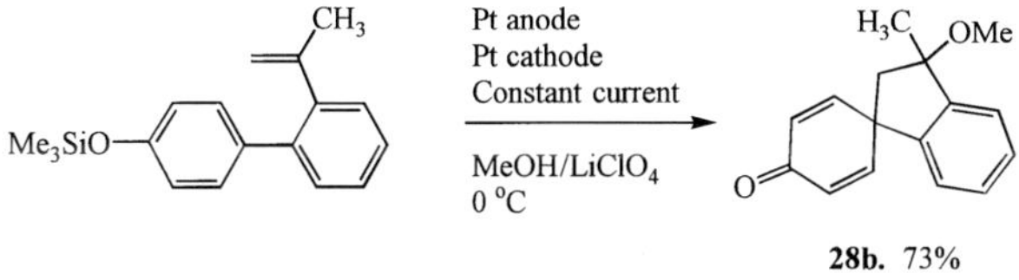

Scheme 31

substituent on the olefin (**29b**). This reaction proceeded in a fashion directly analogous to the substrate not having the methoxy group (**25b**). The anodic oxidation reactions of a pair of substrates having a methoxy group *para* to the phenol and *meta* to the olefin were also studied. In this case, the substrate without the methyl group on the olefin (**29c**) led to none of the desired cyclized product. A 70% yield of a product where methanol had trapped the phenoxonium ion before cyclization was obtained. The anodic oxidation of **29d** again demonstrated the remarkable effect of the methyl substituent on the olefin. In this case, an 80% yield of the cyclized product was obtained.

Evidence was obtained to suggest that the orientation of the oxygen with respect to the aromatic ring was important for the cyclization reaction. The presence of 4,5-dimethoxy substituents on the ring bearing the olefin enhanced the cyclization reaction (Scheme 33). However, the presence of a 4,5-methylenedioxyl group did not enhance the cyclization. This was most clearly observed for the substrates that did not have a methyl substituent on the olefin nucleophile. The use of the 4,5-dimethoxy substituents (**31a**) led to a 70% yield of the cyclized product. The use of the 4,5-methylenedioxyl group (**31b**) led to only 12% of the cyclized material, a result that paralleled the earlier oxidation of **25a**. The substrates bearing a methyl substituent on the olefin both led to cyclized product, although the case using the 4,5-dimethoxy substituents again led to a superior yield.

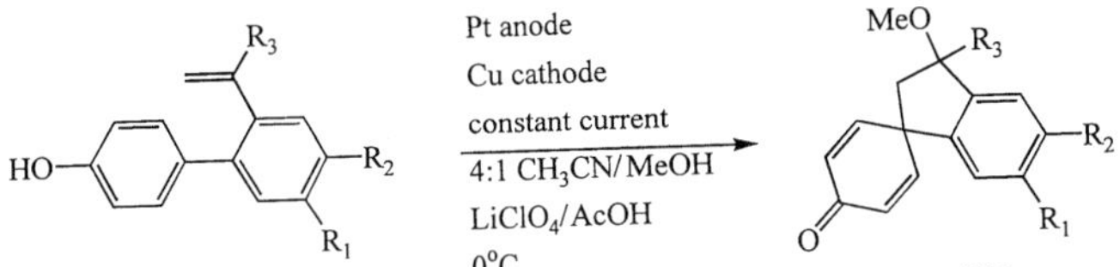

29a. R_1=OMe, R_2=R_3=H
29b. R_1=OMe, R_2=H, R_3=Me
29c. R_1=H, R_2=OMe, R_3=H
29d. R_1=H, R_2=OMe, R_3=Me

30a. 47%
30b. 84%
30c. --
30d. 80%

Scheme 32

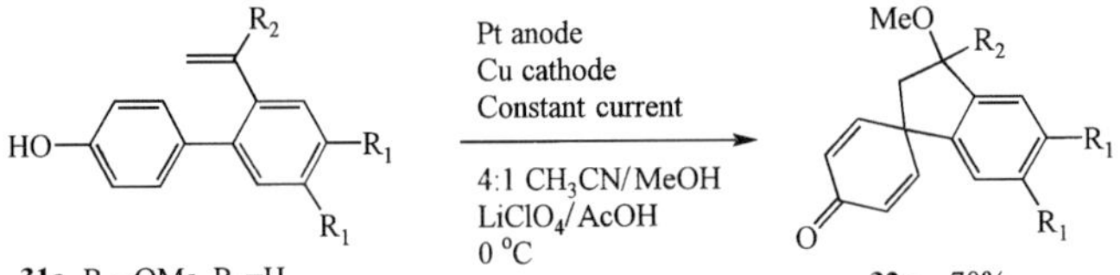

31a. R_1=OMe, R_2=H
31b. R_1=OCH$_2$O, R_2=H
31c. R_1=OMe, R_2=Me
31d. R_1=OCH$_2$O, R_2=Me

32a. 70%
32b. 12%
32c. 92%
32d. 70%

Scheme 33

The chemical oxidation of these substrates (Schemes 28–33) using iodobenzene diacetate was also examined. For the most part, the chemical oxidation and electrochemical oxidation conditions led to similar results.

Cyclization reactions utilizing a vinyl sulfide group were also examined (Scheme 34) [46]. This substrate was chosen for study because, like the methyl substituents used earlier, the sulfide would have a favorable conformational effect on the substrate and would serve as an electron-donating group for making the olefin more nucleophilic. Unlike the methyl substituent, the use of the sulfide led to a carbonyl product that could then be used to further elaborate the cyclized product. Hence, the success of the vinyl sulfide-based cyclization reaction served to extend the synthetic scope of these reactions.

Intramolecular coupling reactions between aryl rings and electron-rich olefins have also been studied as a means for generating fused bicyclic ring skeletons [47]. These reactions were of interest because it was expected that the anodic cyclizations would lead to an addition of the aromatic ring to the olefin with a regiochemistry opposite to what would be expected for the corresponding Friedel-Crafts type of alkylation (Scheme 35). This difference was anticipated because of the anodic oxidation reactions' propensity for reversing the polarity of an enol ether (see below). A successful cyclization reaction would mean that a single substrate could be channeled toward the synthesis of two different ring systems by simply choosing the method by which the cyclization was initiated.

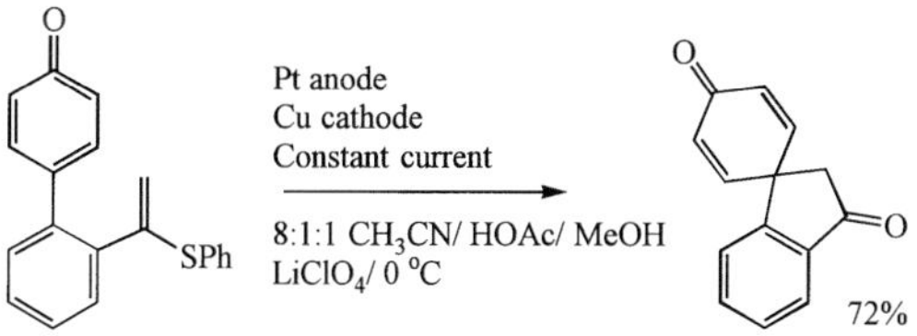

Scheme 34

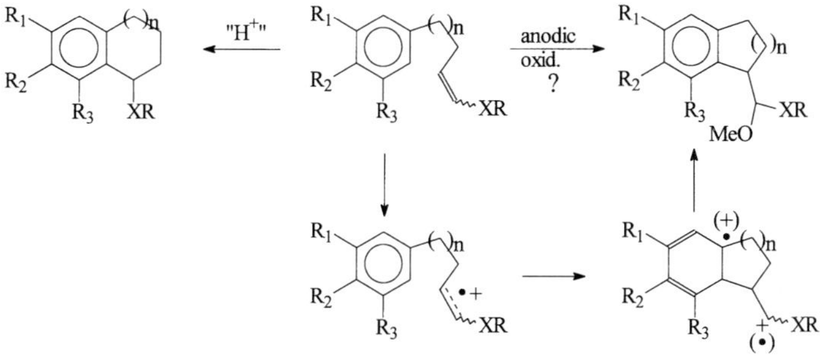

Scheme 35

Although closely related to the reactions of Yamamura and Swenton described above, the anodic coupling reaction outlined in Scheme 35 was significantly different in a number of ways. Foremost among these differences was the fact that in both the chemistry developed by Yamamura and the chemistry developed by Swenton, the reactions led to the formation of a spirocyclic ring skeleton involving the initial phenol ring. The formation of this new ring skeleton was essential because it prevented reformation of the phenol ring and protected the product from over-oxidation. In the anodic cyclization reaction proposed in Scheme 35, no such spirocyclic ring would be formed and the aromatic ring would be regenerated during the reaction. In fact, the newly formed aryl ring would be expected to have a lower oxidation potential than the aryl ring in the starting material. If the aryl ring underwent the initial oxidation reaction, then over-oxidation of the product could not be stopped. Clearly, it was important that the initiating group was the olefin. For this reason, an enol ether was selected as the electron-rich olefin.

The first coupling reaction of this type studied utilized a 3-methoxyphenyl ring as the aryl coupling partner (Scheme 36) [47a, c]. The reaction employed constant current electrolysis conditions and a reticulated vitreous carbon anode (RVC). A good yield of cyclized material was obtained. However, the reaction was plagued by the formation of secondary products derived from over-oxidation (**35** and **36**) of the initially formed cyclization products (**33** and **34**). The amount of over-oxidized material could be greatly reduced with the use of controlled potential electrolysis conditions.

Upon oxidation, a cyclization reaction using a 4-methoxyphenyl ring-derived substrate did not form any of the fused bicyclic product. Instead, a spirocyclic product was formed in direct analogy to the chemistry of Yamamura and Swenton (Scheme 37). In general, cyclization reactions having

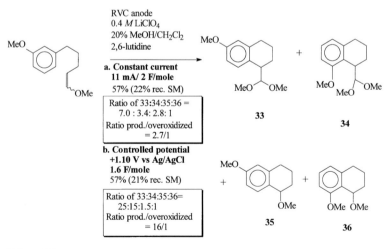

Scheme 36

4-alkoxy substituted phenyl rings were found not to afford synthetically useful amounts of fused bicyclic products.

Anodic coupling reactions using dialkoxy phenyl rings were also studied. When a 3,5-dimethoxy phenyl ring was used as the aryl ring, the coupling reaction benefited strongly from the use of a vinylsulfide initiating group (Scheme 38). When the oxidation of an enol ether was used to initiate the reaction, a 1:1 mixture of cyclized product to over-oxidized material was obtained. However, when the oxidation of the vinyl sulfide was used to initiate the reaction a 72% yield of the desired cyclized product was obtained. None of the over-oxidized material was observed in this reaction.

In addition to the use of phenyl rings, the anodic coupling of electron-rich olefins to furan rings was studied. The goal of this work was to construct bicyclic furan ring skeletons that could be either directly transformed into a variety of bicyclic furan-containing natural products [48] or used in the synthesis of polycyclic natural products by taking advantage of the utility of furans as synthetic intermediates [49]. A wide range of cyclization reactions were studied [47b,c]. A general schematic for these reactions is outlined below in Scheme 39.

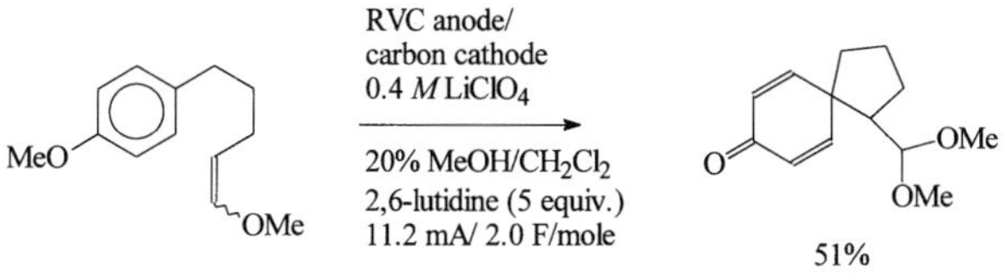

Scheme 37

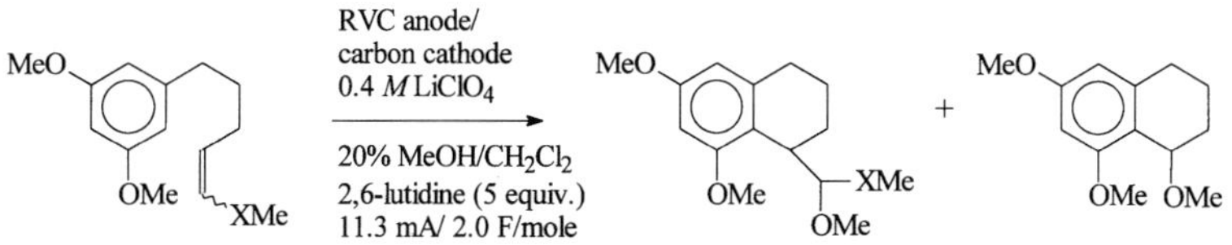

Scheme 38

37a. X = O (28% recovered)
37b. X = S

38a. X=O 31%
38b. X=S 72%

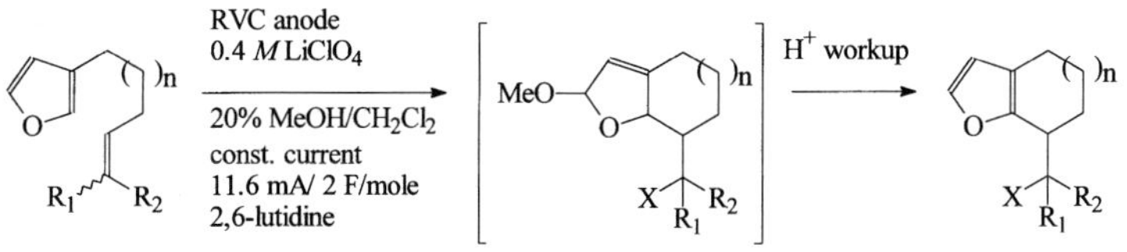

39. 33%
--

Scheme 39

Initially, the reactions were set up to explore six-membered ring formation (n = 1) and triggered by the anodic oxidation of a vinyl sulfide (R_1 = SMe). The reaction led to the formation of the bicyclic **41** which was then converted into the desired bicyclic furan **42** by treatment of the crude reaction mixture with *p*-toluenesulfonic acid. A 71% isolated yield of cyclized products (**42**. X = OMe, R_1 = SMe or OMe) was obtained. The initial formation of **41** was interesting because it indicated that the desired bicyclic furan product did not form during the actual electrolysis step and that there would be no problem with overoxidation of the product. A wide variety of olefins could be used as coupling partners for the furan. To this end, cyclization reactions utilizing enol ethers, styrenes, simple alkyl substitued olefins, and allylsilanes as coupling partners for the furan rings were examined. All of the cyclization reactions studied led to good yields of six-membered ring products, and the use of enol ether and styrene coupling partners were compatible with the formation of seven-membered rings. The coupling reaction between a furan ring and an enol ether to form a six-membered ring was shown to be compatible with the formation of a quaternary carbon.

Finally, the intramolecular coupling reaction between an olefin and a pyrrole ring has been examined (Scheme 40). In this example, a 66% isolated yield of the six-membered ring product was obtained. A vinyl sulfide moiety was used as the olefin participant and the nitrogen protected as the pivaloyl amide in order to minimize the competition between substrate and product oxidation. Unlike the furan cyclizations, the anodic oxidation of the pyrrole-based substrate led mainly to the desired aromatic product without the need for subsequent treatment with acid.

5.2 Intramolecular Anodic Olefin Coupling Reactions

Intramolecular coupling reactions between nucleophilic olefins have also proven to hold potential as synthetically useful reactions. The first example of this type of reaction was reported by Shono and coworkers who examined the intramolecular coupling reaction of an enol acetate and a monosubstituted olefin (Scheme 41) [50]. This reaction was conducted in an effort to probe the nature of the radical cation intermediate generated from the anodic oxidation of

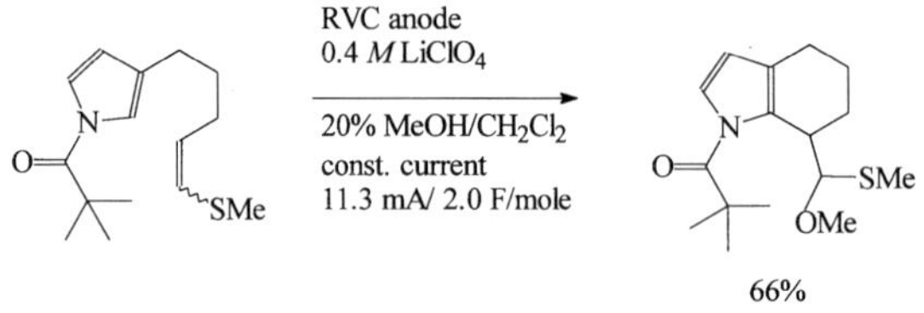

Scheme 40

the enol acetate. A moderate amount of a six-membered ring product was obtained along with hydrolyzed starting material and a small amount of an uncyclized enone. No five-membered ring product was obtained suggesting that the radical cation behaved in an electrophilic fashion. While the yields of cyclized products were low, the reactions were intriguing because of the potential they displayed for reversing the polarity of electron-rich (nucleophilic) olefins and for initiating oxidative cyclization reactions using simple olefin-trapping groups.

The low yield of cyclized product in these first intramolecular anodic olefin-coupling reactions appeared to be due to the conditions used for the reaction. Related intermolecular anodic coupling reactions using enol ethers and stryrenes were known [51]. These reactions worked best when either methanol or acetonitrile solvent was used and often required the use of a base to slow down hydrolysis of the starting enol ether. In order to gain insight into the general applicability of these reactions, a program was initiated to explore the application of the conditions used in the intermolecular coupling reactions to a variety of intramolecular coupling reactions [52]. In these reactions, a methoxy enol ether was used as the initiating group, and a three-carbon tether was used to link the coupling olefins (Scheme 42). While the initial reactions focused on the formation of five-membered rings, both five- and six-membered ring cyclization substrates were routinely examined. Enol ethers, styrenes, simple alkyl substituted olefins, allylsilanes, and vinylsilanes were used as the second coupling partner in the reactions. All of the reactions studied were run

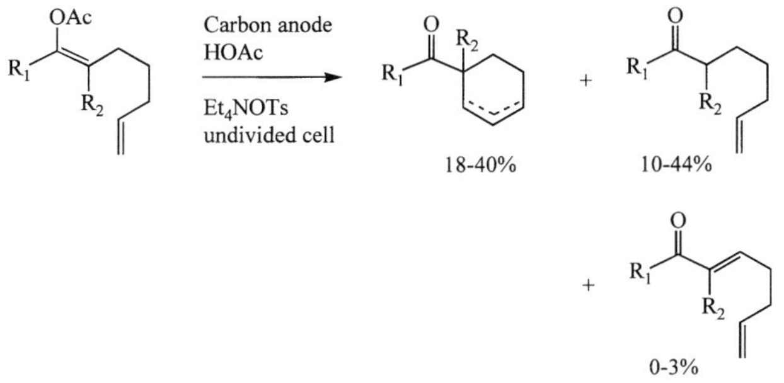

Scheme 41

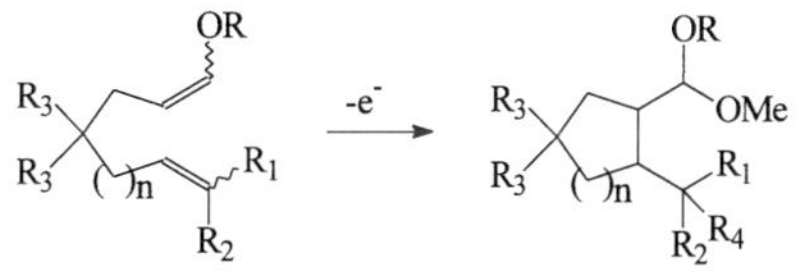

Scheme 42

utilizing constant current conditions at either a carbon or platinum anode. The majority of the reactions used lithium perchlorate as the electrolyte, methanol as a solvent, and 2,6-lutidine as a proton scavenger. In addition, many of the cyclization reactions benefited from the use of a cosolvent such as THF, acetonitrile, or dichloromethane. In these cases, the use of pure methanol as solvent led to the formation of products derived from methanol trapping of the radical cation before cyclization. The most efficient cyclization reactions involved the coupling reaction of an enol ether with a second enol ether and the coupling reaction of an enol ether with a trisubstituted double bond. The reactions using allylsilane terminating groups were particularly intriguing because they led to olefinic products ($R_1 = R_4 = CH_2$) with control over product regiochemistry.

In addition to these initial studies aimed at exploring the synthetic scope of the reactions, an effort was launched to study the overall reactivity of the radical cation intermediates. This work utilized both cyclic voltammetry experiments [52b] and substrates designed to probe the coupling reactions' ability to overcome the steric difficulties associated with the formation of quaternary carbons [53]. Two items from these studies deserve comment here. First, the cyclic voltammetry studies showed a dependence of the substrate oxidation potential on the length of the chain connecting the olefins being coupled. This data was consistent with the cyclization reactions being fast and occurring at or near the anode surface. In addition, this observation provided a method for determining the relative rates of the cyclization reactions. For the simple cyclizations, the reactions were found to follow what would be considered to be a normal reactivity pattern (i.e., five-membered ring cyclizations were faster than six-membered ring cyclizations which were faster than seven-membered ring cyclizations, etc. *gem*-dialkyl substituents increased the rate of the cyclizations). Second, many of the coupling reactions were shown to be compatible with the formation of quaternary carbons (Scheme 43). For example, the coupling of bis enol ether substrates were found to be compatible with the generation of five- and six-membered rings even when carbon-carbon bond formation required construction of a quaternary carbon. The five-membered ring cyclization having $X = OMe$, $n = 1$, and $R = H$ showed a similar drop in potential by cyclic voltammery to that which was observed for the related cyclizations that did not

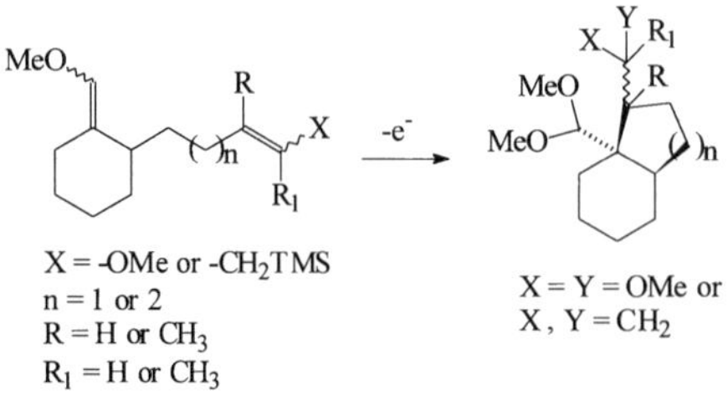

X = -OMe or -CH$_2$TMS
n = 1 or 2
R = H or CH$_3$
R$_1$ = H or CH$_3$

X = Y = OMe or
X, Y = CH$_2$

Scheme 43

involve quaternary carbon formation. Evidently, the cyclization forming the quaternary carbon was still fast and still happened at or near the electrode surface. In addition, the coupling of enol ethers to form five-membered rings was also found to be compatible with the formation of vicinal quaternary carbons (Scheme 43: X = OMe, n = 1, R = CH$_3$). In all cases, the cyclization formed the quaternary carbon in a diastereoselective fashion and led to the fused bicyclic ring skeleton having *cis*-stereochemistry.

Cyclization substrates involving allylsilane groups were not as reactive as their bis enol ether counterparts. While cyclizations utilizing allylsilanes to form five-membered rings and quaternary centers worked well (Scheme 43: X = CH$_2$TMS, n = 1), related cyclizations involving the formation of a six-membered ring and a quaternary center (X = CH$_2$TMS, R = R$_1$ = H, n = 2) were not successful. In this case, the product obtained arose from methanol trapping of the initially formed radical cation. This apparent difference in reactivity between the allylsilane and enol ether groups was consistent with the earlier observation that anodic cyclizations utilizing allylsilanes, like those illustrated in Scheme 42, required the use of a cosolvent, whereas the analogous cyclizations utilizing bis enol ether substrates to form five- and six-membered rings could be accomplished using pure methanol as the solvent.

Throughout these initial studies, the reactions were viewed as if the oxidation reaction led to a radical cation intermediate that then cyclized as if the enol ether group behaved as an electrophile and a cationic intermediate was formed at the terminating end of the cyclization (Scheme 44). This predictive model was based on the work of Shono et al. cited earlier [50]. While other mechanistic scenarios were possible (and perhaps likely), the model for the reaction outlined in Scheme 44 was very successful at predicting and explaining the outcome of the initial intramolecular anodic olefin-coupling reactions studied. For example, it was the observation that the reaction behaved as if it were forming a cation at the terminating end of the cyclization that led to the choice of an allylsilane group for controlling the regiochemistry of product formation.

The success of cyclization reactions utilizing allylsilane groups prompted a study aimed at determining whether vinylsilanes would also be effective for controlling olefin regiochemistry in the cyclization product [54]. To this end, substrate **43** was synthesized and subjected to the anodic electrolysis conditions (Scheme 45). Surprisingly, the oxidation of **43** not only led to the expected six-membered ring product, but also led to the formation of five-membered ring

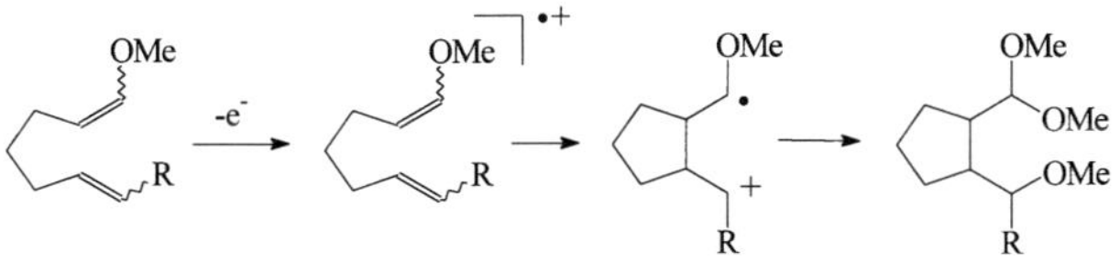

Scheme 44

products. In spite of the low mass balance for the reaction, this was an intriguing result. Based on the predictive model described above, this substrate should have led to the formation of a carbocation beta to the silicon after cyclization, and hence the silane should have *enhanced* the formation of a six-membered ring product. The directly analogous cyclization without the silyl group on the terminating olefin afforded exclusively 6-membered ring products. Clearly, something was wrong with the predictive model outlined in Scheme 44.

The results with the vinyl silane substrate were more consistent with an oxidation that led to a radical at the terminal end, especially if the initial cyclization was viewed as being reversible (Scheme 46). In this way, when R in

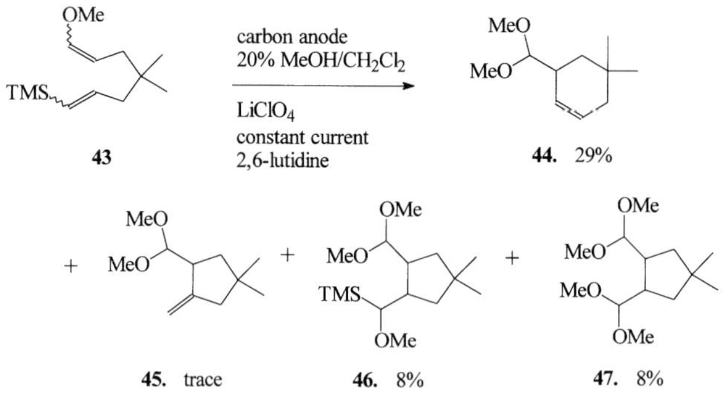

Scheme 45

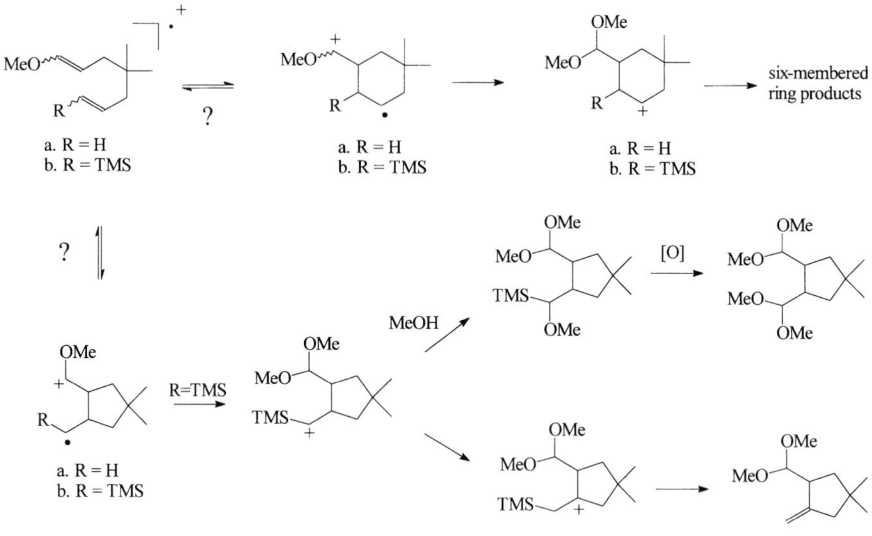

Scheme 46

Scheme 46 was equal to H the cyclization leading to the five-membered ring product would lead to a primary radical that could not be oxidized further to a cation. The reaction would reverse to the starting radical cation and then give rise to a cyclization affording the six-membered ring product having a secondary radical for subsequent oxidation. However, when R was equal to a TMS group, the cyclization leading to the five-membered ring product would be capable of undergoing a second oxidation. In this scenario, the reaction would lead to a mixture of five- and six-membered ring products as observed.

Further evidence for this mechanistic picture was obtained by looking at the oxidation of substrate **48** (Scheme 47). Recent work indicating that radical cyclizations involving silyl-based terminating groups show a preference for the formation of α-silyl radicals over β-silyl radicals [55] and the well known preference for cationic cyclizations involving vinylsilanes to afford β-silyl cations preferentially over α-silyl cations suggested that the oxidation of substrate **48** might shed light on the nature of the radical cation cyclization. If the cyclization led to products having a radical intermediate on the terminating end and an oxonium ion on the initiating end, then the presence of the silyl would be expected to lead to an increased preference for six-membered ring formation. If the opposite scenario were to govern the reaction, then the presence of the silyl group would be expected to lead to an increase in five-membered ring formation. For comparison, the oxidation of the directly analogous substrate without the silyl group led to a mixture of five- (27% yield) and six-membered ring (9% yield) products. The anodic oxidation of **48** led to the formation of exclusively six-membered ring products (58% yield). No five-membered ring product could be observed in the 300 MHz ^{1}H NMR of the crude reaction. Hence, the reaction was consistent with a model that had a radical intermediate at the terminating end of the cyclization.

The relatively low mass balances of the reactions with vinylsilane terminating groups raised questions about the overall utility of vinylsilane groups with the anodic olefin-coupling reaction. Were the low mass balances due to an

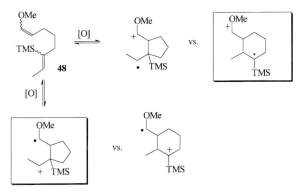

Scheme 47

inherent problem involving the compatibility of the vinylsilane with the electrolysis conditions, or was the low mass balance due to a problem in a subsequent step of the reaction? In order to address this question, a substrate was built along the lines of an empirical observation that had been made concerning past cyclizations. When a cyclization substrate was predicted to afford the same product regardless of whether it proceeded through a kinetic radical type cyclization or a cationic type cyclization, it always led to a good yield of product. For the use of a vinylsilane in these cyclizations, compound **49** represented such an ideal substrate (Scheme 48) [53, 54]. If the initial oxidation occured at the enol ether, then either a kinetic radical-type cyclization or a cationic-type cyclization would be expected to lead to five-membered ring formation. In practice, the anodic oxidation of **49** led to the formation of a 72% isolated yield of product where the vinyl group had been transferred from the silyl moiety to the β-carbon of the enol ether. This result was exciting because it showed that vinylsilanes were compatible with the electrolysis conditions, illustrated the potential for electrochemistry as a tool for synthesizing quaternary centers with control of relative stereochemistry, and demonstrated the utility of electrochemistry for generating radical cations from *very* acid sensitive substrates.

Having established that anodic olefin-coupling reactions could efficiently lead to carbon-carbon bond formation, attention was turned toward developing synthetic stategies that would take advantage of the resulting functionalized products. For example, the cyclizations illustrated in Scheme 43 afforded bicyclic 1,4-dicarbonyl equivalents that in principle could serve as potential substrates for a subsequent intramolecular aldol condensation [56]. Such a sequence would afford angularly fused tricyclic enone building blocks.

The most direct route to the 1,4-dicarbonyl equivalent required for the aldol condensation would be to couple the enol ether of an aldehyde with the enol ether of a ketone. However, this sequence proved impractical due to the hydrolytic instability of the ketone enol ether. Even after an extensive effort, the substrate for the electrolysis reaction could not be reproducibly prepared in high yield. These problems were readily avoided with the use of an allylsilane based

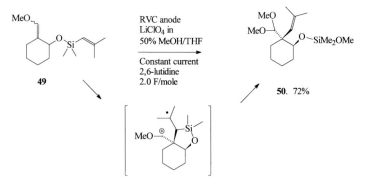

Scheme 48

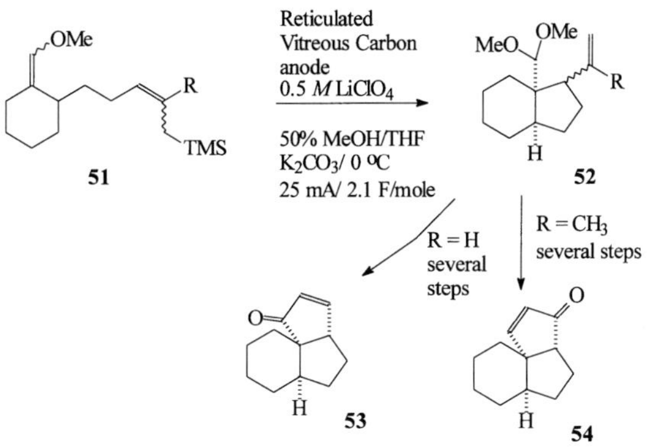

Scheme 49

substrate (Scheme 49). Ozonolysis of the resulting olefinic product then afforded the desired 1,4-dicarbonyl equivalent. In this way, a tandem electrochemical oxidation-aldol condensation strategy was developed for the generation of angularly-fused tricyclic enone building blocks **53** and **54**.

6 Conclusions

Over the past ten years, intramolecular anodic C–C bond-forming reactions have steadily grown in terms of the number and types of transformations reported. These unique cyclization reactions have led to products ranging from simple monocyclic ring systems to complex bridged and fused polycyclic ring skeletons. In many of the more synthetically applicable examples, the oxidative nature of the electrolysis reaction allowed for the formation of cyclic products while maintaining the functional groups used to initiate and terminate the reaction. The high level of functionality in these products appears to be ideal for further synthetic elaboration. However, in spite of the numerous successes reported and the apparent potential for future development, many questions concerning the overall synthetic utility of anodic cyclization reactions still remain. Are intramolecular anodic C–C bond-forming reactions really a generally useful tool for constructing complex organic molecules? If so, then what new retrosynthetic pathways do they make available, and how do these retrosynthetic pathways complement already existing chemical routes? Answers to questions like these need to be addressed if anodic cyclization reactions are to become a generally accepted technique by the larger synthetic community. Over the next ten years, it should be interesting to watch how the chemistry reviewed here is applied to this challenge.

7 References

1. For a general overview of organic electrochemistry see a) Lund H, Baizer MM (1991) Organic electrochemistry: an introduction and a guide, 3rd ed. M Dekker, New York. For overviews of anodic electrochemistry see b) Torii S (1985) Electroorganic synthesis: methods and applications: part I – oxidations. VCH, Deerfield Beach, FL; c) Yoshida K (1984) Electrooxidation in organic chemistry: the role of cation radicals as synthetic intermediates. John Wiley and Sons, New York; d) Ross SD, Finkelstein M, Rudd EJ (1975) Anodic oxidation. Academic Press, New York; e) Schäfer HJ (1981) Angew Chem Int Ed Eng 20: 911
2. For excellent summaries of recent work see a) Torii S (ed) (1995) Novel trends in electroorganic synthesis. Kodansha, Tokyo; b) Nonaka T, Tokuda K (eds) (1994) Denki Kagaku, 62 (12); c) Little RD, Weinberg NL (1991) Electroorganic synthesis: Festschrift for Manuel M. Baizer, M Dekker, New York; d) Fry A (1993) Aldrichimica Acta 26: 3; e) Swenton JS, Morrow GW (eds) (1991) Synthetic applications of anodic oxidations, Tetrahedron 42
3. For reactions using $Mn(OAc)_3$ see a) Snider BB (1996) Chem Rev 96:339; b) Snider BB, Cole BM (1995) J Org Chem 60: 5376; c) Zhang Q, Mohan R, Cook L, Kazanis S, Peisach D, Foxman BM, Snider BB (1993) J Org Chem 58: 7640; d) Snider BB, Wan BYF, Buckman BO, Foxman BM (1991) J Org Chem 56: 328; e) Snider BB, Merritt JE, Dombroski MA, Buckman BO (1991) J Org Chem 56: 5544; f) Curran DP, Morgan TM, Schwartz CE, Snider BB, Dombroski MA (1991) J Am Chem Soc 113: 6607; g) Dombroski MA, Kates SA, Snider BB (1990) J Am Chem Soc 112: 2759; h) Kates SA, Dombroski MA, Snider BB (1990) J Org Chem 55: 2427; i) Snider BB, Patricia JJ (1989) J Org Chem 54: 38; j) Snider BB, Patricia JJ, Kates SA (1988) J Org Chem 53: 2137; k) Snider BB, Dombroski MA (1987) J Org Chem 52: 5487; l) Mohan R, Kates SA, Dombroski MA, Snider BB (1987) Tetrahedron Lett 28: 845; m) Snider BB, Mohan R, Kates SA (1987) Tetrahedron Lett 28: 841; n) Snider BB, Mohan R, Kates SA (1985) J Org Chem 50: 3659; o) Corey EJ, Ghosh AK (1987) Tetrahedron Lett 28: 175; p) Corey EJ, Gross AW (1985) Tetrahedron Lett. 26: 4291; q) Corey EJ, Kang M (1984) J Am Chem Soc 106: 5384; r) Fristad WE, Peterson JR, Ernst AB, Urbi GB (1986) Tetrahedron 42: 3429; s) Fristad WE, Hershberger SS (1985) J Org Chem 50: 1026; t) Fristad WE, Peterson JR, Ernst AB (1985) J Org Chem 50: 3143; u) Fristad WE, Peterson JR (1985) J Org Chem 50:10; v) Fristad WE, Ernst AB (1985) Tetrahedron Lett 26: 3761. For reactions using other metals see w) Pattenden G, Bhandal H, Russell JJ (1986) Tetrahedron Lett 27: 2299; x) Pattenden G, Patel VF, Russell JJ (1986) Tetrahedron Lett 27: 2303; y) Kraus GA, Landgrebe K (1984) Tetrahedron Lett 25: 3939; z) Baldwin JE, Li CS (1987) J Chem Soc, Chem Commun 1987: 166
4. Snider BB, Kwon T (1990) J Org Chem 55: 4786
5. For a recent review of the Kolbe electrolysis reaction see Schäfer HJ (1990) Topics in Current Chemistry 152: 91
6. For an excellent review of aryl–aryl coupling reactions see reference 1c, section 4-3 and references therein
7. Huhtasaari M, Schäfer HJ, Becking L (1984) Angew Chem Int Ed Eng 23: 980
8. Miller LL, Stermitz FR, Falck JR (1973) J Am Chem Soc 95: 2651
9. Tabakovic I (1996) Topics in Current Chemistry, this volume.
10. For pioneering work with anodic amide oxidations see: a) Ross SD, Finkelstein M, Peterson C, (1964) J Am Chem Soc 86: 4139; b) Ross SD, Finkelstein M, Peterson C (1966) J Org Chem 31: 128, and c) Ross SD, Finkelstein M, Peterson, C (1966) J Am Chem Soc 88: 4657. For reviews of anodic amide oxidations see d) Shono T (1984) Tetrahedron 40: 811 and e) Shono T, Matsumura Y, Tsubata K (1984) Organic Synthesis 63: 206
11. Shono T, Matsumura Y, Tsubata K (1981) J Am Chem Soc 103: 1172
12. Shono T (1988) Topics in Current Chemistry 148: 131
13. a) Shono T, Matsumura Y, Inoue K (1983) J Org Chem 48: 1388; b) Thaning M, Wistrand LG (1986) Helv Chim Acta 69: 1711; c) Renaud P, Seebach D (1986) Angew Chem Int Ed Eng 25: 843; d) Renaud P, Seebach D (1986) Helv Chim Acta 69: 1704; e) Seebach D, Charczuk R, Gerber C, Renaud P, Berner H, Schneider H (1989) Helv Chim Acta 72: 401; f) Papadopoulos A, Heyer J, Ginzel KD, Steckhan E (1989) Chem Ber 122: 2159; g) Ginzel KD, Brungs P, Steckhan E (1989) Tetrahedron 45: 1691; g) Papadopoulos A, Lewall B, Steckhan E, Ginzel KD, Knoch F, Nieger M (1991) Tetrahedron 47: 563; i) Barrett AGM, Pilipauskas D (1991) J Org Chem 56: 2787; j) Moeller KD, Rothfus SL, Wong PL (1991) Tetrahedron 47: 583; k) Moeller KD,

Rutledge LD (1992) J Org Chem 57: 6360; l) Cornille F, Fobian YM, Slomczynska U, Beusen DD, Marshall GR, Moeller KD (1994) Tetrahedron Lett 35: 6989; m) Cornille F, Slomczynska U, Smythe ML, Beusen DD, Moeller KD, Marshall GR (1995) J Am Chem Soc 117: 909; n) Slomczynska U, Chalmers DK, Cornille F, Smythe ML, Beusen DD, Moeller KD, Marshall GR (1996) J Org Chem 61: 1198, and o) Fobian YM, d'Avignon DA, Moeller KD (1996) Bioorg Med Chem Lett 6: 315

14. Wong PL, Moeller KD (1993) J Am Chem Soc 115: 11434
15. Li W, Hanau CE, d'Avignon A, Moeller KD (1995) J Org Chem 60: 8155
16. Unpublished results with Li W, Gershengorn M, Perlman J, and Moeller KD. For related work with conformationally restricted pyroglutamate rings see: Rutledge LD, Perlman JH, Gershengorn MC, Marshall GR, Moeller KD (1996) J Med Chem 39: 1571
17. For reviews of chemical approaches to radical cyclization reactions see: a) Motherwell WB, Crich D (1992) Free radical chain reactions in organic synthesis, Academic Press, London, b) Giese B (1986) Radicals in organic synthesis: formation of carbon–carbon bonds, Pergamon, Oxford, and c) Jasperse CP, Curran DP, Fevig TL (1991) Chem Rev 91: 1237
18. Becking L, Schfer HJ (1988) Tetrahedron Lett 29: 2797
19. Weiguny J, Schäfer HJ (1994) Liebigs Ann Chem 235
20. Kishi M, Takahashi K, Kawanda K, Goh Y (1991) EP 0471856A1 and (1992) Chem Abst 116: P59072s
21. For chemical routes to tandem radical cyclization reaction see Curran DP, Kuo SC (1987) Tetrahedron 43: 5653, as well as reference 17c.
22. Matzeit A, Schäfer HJ (1995) Radical tandem cyclizations. In: Torii S (ed) Novel trends in electroorganic synthesis, Kodansha, Tokyo
23. Yoshida J, Sakaguchi K, Isoe S (1988) J Org Chem 53: 2525
24. Snider BB, McCarthy BA (1994) ACS symposium series: benign by design 577: 84
25. Shundo R, Nishiguchi I, Matsubara, Y Hirashima, T (1991) Chem Lett 1991: 235
26. Shundo R, Nishiguchi I, Matsubara Y, Hirashima T (1991) Tetrahedron 47: 831
27. Nédélec JY, Nohair K (1991) Synlett 1991: 659
28. Bergamini F, Citterio A, Gatti N, Nicolini M, Santi R, Sebastiano R (1993) J Chem Research (S) 1993: 364
29. Warsinsky R, Steckhan E (1995) Oxidative free radical additions of α-nitro amides to alkenes and alkynes mediated by electrochemically regenerable manganese (III) acetate. In: Torii S (ed) Novel trends in electroorganic synthesis, Kodansha, Tokyo, pg 135; J. Chem. Soc. Perkin Trans 1 (1994) 1994: 2027
30. Elinson MN, Feducovich SK, Lizunova TL, Nikishin GI (1995) Electroorganic synthesis of substituted cyclopropanes. In: Torii S (ed) Novel trends in electroorganic synthesis, Kodansha, Tokyo, pg 47
31. Yoshida J, Ishichi Y, Isoe S (1992) J Am Chem Soc 114: 7594
32. Yoshida J, Takada K, Ishichi Y, Isoe S (1995) Electrooxidative cyclization using group 14 metals. In: Torii, S (ed) Novel trends in electroorganic synthesis, Kodansha, Tokyo, pg 295
33. Yoshida J, Maekawa T, Murata T, Matsunaga S, Isoe S (1990) J Am Chem Soc 112: 1962
34. Ohmori H, Maki T, Maeda H (1995) Synthetic application of acyl- and alkoxy-phosphonium ions. In: Torii, S (ed) Novel trends in electroorganic synthesis, Kodansha, Tokyo, pg 345
35. Maeda H, Maki T, Ohmori H (1992) Tetrahedron Lett 33: 1347
36. Yamamura S (1995) Synthetic studies on natural products using electrochemical method as a key step. In: Torii, S (ed) Novel trends in electroorganic synthesis, Kodansha, Tokyo, pg 265
37. Maki S, Kosemura S, Yamamura S, Kawano S, Ohba S (1992) Chem Lett 1992: 651
38. Maki S, Toyoda K, Kosemura S, Yamamura S (1993) Chem Lett 1993: 1059
39. Maki S, Asaba N, Kosemura S, Yamamura S (1992) Tetrahedron Lett 33: 4169
40. Shizuri Y, Okuno Y, Shigemori H, Yamamura S (1987) Tetrahedron Lett 28: 6661
41. Maki S, Kosemura S, Yamamura S, Ohba S (1993) Tetrahedron Lett 34: 6083
42. Shizuri Y, Maki S, Ohkubo M, Yamamura S (1990) Tetrahedron Lett 31: 7167
43. Maki S, Toyoda K, Mori T, Kosemura S, Yamamura S (1994) Tetrahedron Lett 35: 4817
44. a) Morrow GW, Chen Y, Swenton JS (1991) Tetrahedron 47: 655 and b) Morrow GW, Swenton JS (1987) Tetrahedron Lett 28: 5445
45. Swenton JS, Carpenter K, Chen Y, Kerns ML, Morrow GW (1993) J Org Chem 58: 3308
46. Swenton JS, Callinan A, Chen Y, Rohde JJ, Kerns ML, Morrow GW (1996) J Org Chem 61: 1267

47. a) Moeller KD, New DG (1994) Tetrahedron Lett 35: 2857; b) Tesfai Z, Moeller KD (1994) Denki Kagaku 62: 1115, and c) New DG, Tesfai Z, Moeller KD (1996) J Org Chem 61: 1578
48. a) Danheiser RL, Stoner EJ, Koyama H, Yamashita DS, Klade C (1989) J Am Chem Soc 111: 4407; b) Carté B, Kernan MR, Barrabee EB, Faulkner DJ (1986) J Org Chem 51: 3528; c) Gopalan A, Magnus P (1984) J Org Chem 49: 2317; d) Tanis SP, Dixon LA (1987) Tetrahedron Lett 28: 2495; e) Hiroi K, Sato H (1987) Synthesis 1987: 811, and f) Padwa A, Ishida M (1991) Tetrahedron Lett 32: 5673
49. For reviews see a) Dean FM (1982) Adv. Heterocycl. Chem 30: 161 and b) Lipshutz, BH (1986) Chem Rev 86: 795
50. Shono T, Nishiguchi I, Kashimura S, Okawa M (1978) Bull Chem Soc Jpn 51: 2181
51. a) Belleau B, Au-Young YK (1969) Can J Chem 47: 2117; b) Fritsch JM, Weingarten H (1968) J Am Chem Soc 90: 793; c) Fritsch JM, Weingarten H, Wilson JD (1970) J Am Chem Soc 92: 4038; d) Le Moing MA, Le Guillanton G, Simonet J (1981) Electrochim Acta 26: 139; e) Engels R, Schäfer HJ, Steckhan E (1977) Liebigs Ann Chem 1977: 204; f) Schäfer HJ, Steckhan E (1970) Tetrahedron Lett 11: 3835; g) Schäfer HJ, Steckhan E (1974) Angew Chem Int Ed Engl 13: 472; h) Koch D, Schäfer HJ, Steckhan E (1974) Chem Ber 107: 3640 and i) Fox MA, Akaba R (1983) J Am Chem Soc 103: 3460
52. For an overview of early work see a) Hudson CM, Marzabadi MR, Moeller KD, New DG (1991) J Am Chem Soc 113: 7372 and b) Moeller KD, Tinao LV (1992) J Am Chem Soc 114: 1033
53. Moeller KD, Hudson CM, Tinao-Wooldridge LV (1993) J Org Chem 58: 3478
54. Hudson CM, Moeller KD (1994) J Am Chem Soc 116: 3347
55. Miura K, Oshima K, Utimoto K (1989) Tetrahedron Lett 30: 4413
56. Tinao-Wooldridge LV, Moeller KD, Hudson CM (1994) J Org Chem 59: 2381

Anodic Synthesis of Heterocyclic Compounds

Ibro Tabaković

Department of Chemistry, University of Minnesota, Minneapolis MN 55455, USA

Table of Contents

1 Introduction	88
2 Mechanistic Considerations	89
2.1 Intramolecular Cyclizations	89
2.2 Intermolecular Cyclizations	95
2.3 Heterogeneous vs Homogeneous Oxidation	102
3 Anodic Synthesis of Heterocycles	108
3.1 Formation of the Carbon–Nitrogen Bond	108
3.1.1 Oxidation of Amines and Hydroxyamines	108
3.1.2 Oxidation of Hydrazones and Schiff Bases	110
3.1.3 Oxidation of Imidamides	112
3.1.4 Oxidation of Hetero-Allenes	114
3.1.5 Oxidation of Amides	117
3.1.6 Oxidation of Activated Aromatic Rings	118
3.2 Formation of the Carbon–Oxygen Bond	119
3.2.1 Oxidation of Endiamines and Enaminones	119
3.2.2 Oxidation of N-Acylhydrazones and 1-Arylmethylenesemicarbazides	121
3.2.3 Oxidation of Enols and Olefins	122
3.2.4 Oxidation of Alcohols and Catechol	125
3.2.5 Oxidation of Amides and Sulfides	129
3.3 Formation of the Carbon–Sulfur Bond	131
3.4 Formation of the Nitrogen–Nitrogen Bond	131
3.4.1 Oxidation of Hydrazones and Formazans	131
3.4.2 Oxidation of Triazenes	132

3.5 Formation of the Nitrogen–Oxygen Bond 132
    3.5.1 Oxidation of Vicinal Dioximes 132
3.6 Formation of the Nitrogen–Sulfur Bond 136
    3.6.1 Oxidation of Thioamides 136

4 References 136

This article discusses the anodic synthesis of heterocyclic compounds that have appeared during the last decade. The mechanistic aspects involving intramolecular, intermolecular cyclizations and the homogeneous vs heterogeneous anodic oxidations were considered. This review deals with the recent advances in anodic oxidations in which heterocyclic compounds were synthesized through carbon-heteroatom and heteroatom-heteroatom bond formation.

1 Introduction

Organic electrosynthesis is not a new technique and there is considerable knowledge of the types of reactions which take place at cathodes and anodes [1]. One view of organic electrochemistry is that it is a unique non-thermal method for activating molecules [2]. Since the rate of reaction can normally be increased by raising the electrode potential, it is possible to carry out reactions with a high activation energy at low temperature. Another view is that electrochemistry is an alternative to chemical redox methods. Indeed, in certain cases, the products are similar. This is to be expected if the chemical reagent reacts like an electrode via discrete electron transfer steps – not atom transfers. However, it is not unusual to observe significant differences between electrochemical and chemical processes. A particular advantage of electrochemistry is control of the electrode potential. In particular one can adjust the potential to attack selectively the most easily oxidized moiety in a complex molecule. This technique can also avoid the overoxidations produced by chemicals. By using various electroanalytical techniques the mechanism of the redox processes can be studied in detail. The most informative technique in electroanalytical chemistry is cyclic voltammetry which is described as "electrochemical spectroscopy" [3].

The use of the electrode as a valuable tool for synthesis of heterocyclic compounds has been known for many years [1]. There are several good reviews [4] on different aspects of the electrochemistry of heterocyclic compounds. The aim of this article is to cover the important developments during the last decade, as well as some older works that have not been previously reviewed. This review considers the anodic reactions in which a heterocyclic compound is formed through carbon-heteroatom and heteroatom-heteroatom bond formation. The synthesis of heterocycles which occur through carbon–carbon bond formation are not included in the present review.

2 Mechanistic Considerations

2.1 Intramolecular Cyclizations

Intramolecular cyclizations are processes in which the anodic oxidation of a neutral molecule, with two functional groups connected by a tether of atoms, results in ring formation through the loss of two electrons and two protons and a subsequent cyclization step. The electron loss from one of terminal functional groups induces an inversion of the polarity which gives rise to an open chain radical cation which than acts as both electrophile and acid. If the length and nature of the chain (tether) linking terminal groups attains the proper geometry the cyclization reaction takes place between a nucleophilic site and electrophilic site (Scheme 1).

The primary mechanistic objective is the determination of the sequence of the five reaction steps and the rate-determining step. Saveant and coworkers developed a powerful method for mechanistic analysis of cyclization reactions by deriving wave equations and carrying out a large number of digital computer simulations for linear sweep voltammetry (LSV), rotating disc electrode voltammetry, classical polarography [5] and convolution potential sweep voltammetry (CPSV) [6]. It was assumed that the cyclization reaction rate is such that the initial electron transfer remains Nernstian and that the cyclization process corresponds to a two-electron overall exchange along a single wave. The variation of the peak potential (Ep) in LSV over a wide range of sweep rates (v), the initial concentration of the substrate and base provide a means of discriminating between various possible mechanisms in an anodic cyclization reaction. In Scheme 2, only two reaction sequences, each of them consisting of five reaction steps, are shown with the purpose of illustrating the existing formalism. One reaction sequence represents a possible mechanisms involving a cyclization of the initially formed radical cation, and the second represents cyclization of the dication produced by further oxidation of the radical cation at the electrode or by a solution electron transfer. Saveant described any particular mechanism by a combination of five letters, one of which is capitalized to represent the rate-determining step. The heterogeneous transfer from the substrate to the electrode is represented by an "e", the deprotonation step as a "p", and the cyclization step as a "c". The second electron can be transferred in solution through the disproportionation reaction and this reaction step is represented by

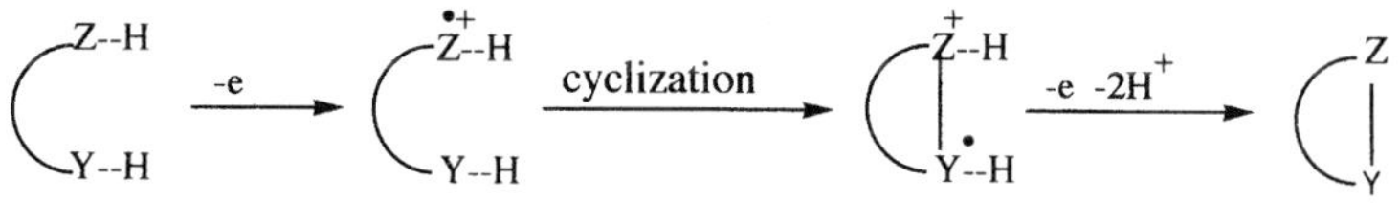

Scheme 1.

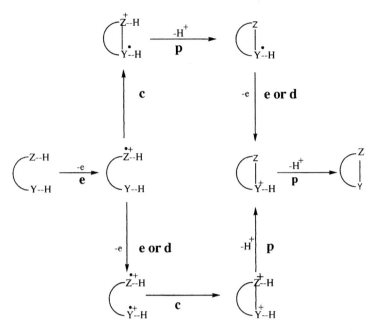

Scheme 2.

a "**d**". The disproportionation equilibrium, $K_d = \exp\{(-F/RT)\Delta E^0\}$ should give a significant concentration of the product of the second electron transfer if ΔE^0 is not too large. The open chain radical cation or cyclized radical cation are intrinsically very acidic. It has been shown recently that pKa (RH$^{·+}$/R$^·$ + H$^+$) of the radical cation of NADH analogs in CH$_3$CN range from -1.9 to 4.7 [7]. The radical cations are easily deprotonated by the base present, thus giving rise to the corresponding radicals which can be oxidized at the anode (**e**-step) or act as electron donors and transfer the second electron in the solution to another radical cation, as an electron acceptor (**d**-step). It should be pointed out that experience has shown that in almost all cases the second electron transfer is of the homogeneous type [8].

If the first **e** step, i.e., heterogeneous electron transfer, is slow (non-Nernstian) or if the cyclization reaction is faster than the electron transfer itself, the electron transfer becomes rate-determining and nothing can be done about the mechanism of cyclization.

The experimental mechanistic study of the anodic cyclization reactions requires values for the variation of the peak potential (Ep) in LSV with the sweep rate (v), the concentration of substrate (C) and the concentration of added base (B). The plots of dEp/dlog v, dEp/dlog C, and dEp/dlog B provide an effective tool for qualitative mechanistic analysis. The diagnostic criteria developed for discrimination between the various possible mechanisms [5] and adopted for oxidative cyclizations are presented in Table 1.

Table 1. Diagnostic criteria for representative anodic cyclizations

Mechanism	dEp/dlog v (mV/decade)	− dEp/dlog C (mV/decade)	− dEp/dlog B (mV/decade)
e-c-e-P-e	14.8	0	14.8
e-C-e-p-p	29.6	0	0
e-c-d-P-p	19.7	0	19.7
e-c-D-p-p	19.7	19.7	0
e-C-d-p-p	29.6	0	0
e-c-P-e-p	29.6	0	29.6
e-C-p-e-p	29.6	0	0
e-c-p-D-p	19.7	19.7	19.7
e-c-P-d-p	29.6	0	29.6
e-C-p-d-p	29.6	0	0
e-e-c-P-p	14.8	0	14.8
e-e-C-p-p	14.8	0	0
e-d-c-P-p	19.7	0	19.7
e-d-C-p-p	19.7	0	0
e-D-c-p-p	19.7	19.7	0

It is certain that all possible mechanistic schemes are not covered in Table 1. Recently this problem was discussed by Fry et al. [9]. Namely, the number of possible permutations of two heterogeneous electron transfers, two proton transfers, and a cyclization step is given by 5!/2! 2! (= 30). Each of these 30 paths has a counterpart in which the second electron transfer step is that of type **d** which result in 60 permutations. That each of the five steps might be rate-determining implies the existence of 300 discrete mechanisms. In spite of the complexity, LSV can be considered as an important method for mechanistic analysis using diagnostic criteria developed by Andrieux and Saveant [5] as well as empirical rules developed by Parker [10]. It is possible to reduce the number of theoretical mechanistic sequences by taking into account the simple chemical arguments. For example, most of the anodic cyclization reactions are initiated by heterogeneous electron transfer and the sequence is then reduced to the four remaining steps. The arguments obtained from other methods [11, 12] (e.g., spectroscopy, kinetic and isotope effects, product analysis) could give further insight into the reaction mechanism. This should serve as a reminder that the mechanisms are correct only as long as they account for all the experimental facts [13]. Data obtained by LSV measurements should be treated carefully for three reasons. First of all, the systems do not give ideal response and voltammograms do not have the theoretical shapes necessary for analysis due to the adsorption and/or "filming" of the electrode. Second, the observation of non-integral reaction orders (e.g., between 19.7 and 29.6 mV) is the reason that the process consists of competing reaction mechanisms [12]. Third, and most important, is the precision in measuring the peak potentials. Parker and coworkers reported that Ep can be measured to ± 0.1 mV using the application of derivative readout techniques to CV [14], or that Ep/2 can be measured with a precision of ± 1 mV using X–Y recording [15]. Recently, Saveant and coworkers have developed the method for measuring Ep with a precision of 0.3–0.4 mV [16].

In order to illustrate the application of LSV in mechanistic analysis we can look at the redox behavior of the formazan-tetrazolium salt system which we studied some years ago [17]. 1,3,5-Triphenyl formazane was oxidized at controlled potential in CH_3CN-Et_4NClO_4 solution to 2,3,5-triphenyl tetrazolium perchlorate which was then isolated in quantitative yield. Coulometry showed that the overall electrode reaction was a two-electron oxidation. It has been shown that the rate of variation of Ep with log v was 30 mV per decade of sweep rate and that there was no variation of the peak potential with the concentration of 1,3,5-triphenylformazan. According to Saveant's diagnostic criteria (Table 1), four mechanistic schemes were possible: **e-C-e-p-p**, **e-C-d-p-p**, **e-c-P-e-p** and **e-c-P-d-p**. If cyclization is the rate-determining step, then the resulting **e-C-e-p-p** and **e-C-d-p-p** mechanisms would not imply variation of Ep with the concentration of base. However, we have observed the 35 mV shift of Ep cathodically in the presence of 4-cyanopyridine as a base. These observations ruled out the first two mechanisms. The remaining possibilities were then **e-c-P-e** and **e-c-P-d**, as shown in Scheme 3.

Our final conclusion was therefore that the mechanism of intramolecular oxidative cyclization of 1,3,5-triphenylformazan to 2,3,5-triphenyltetrazolium perchlorate involves cyclization of the initial radical cation and deprotonation as the rate-determining step, following the mechanistic scheme **e-c-P-e**.

The reduction of tetrazolium salt to formazan was found to occur by opposite mechanistic pathways, i.e., through **e-P-c-e** mechanism from those for electrochemical oxidation of 1,3,5-triphenyformazan to 2,3,5-triphenyltetrazolium salt. The formalism in this mechanistic scheme implies that **c** is a fast ring-opening reaction following the slow protonation reaction. The cyclic voltammogram of the tetrazolium salt run in CH_3CN-0.1 mol/l Et_4NClO_4 solution in the presence of activated alumina showed a reversible, one electron oxidation wave at $E^0 = -0.49$ V vs SCE indicating the formation of a stable tetrazolinyl radical without the presence of proton donor. Our findings were supported later by the result obtained via electrochemical reduction of tetrazolium salts [18, 19].

The discrimination between **e-c-P-e** (Scheme 3.) and **e-c-P-d** mechanisms requires the answer to the question of whether the second electron is transferred through heterogeneous electron transfer (**e**-step) or through solution electron transfer (**d**-step). The following solution electron transfers (Eqs. 1–3) could be considered:

$$RH^{·+} + cR^· \rightleftharpoons RH + cR^+ \tag{1}$$

$$2cR^· \rightleftharpoons cR^+ + R^- \tag{2}$$

$$RH^{·+} + cRH^{·+} \rightleftharpoons RH + cRH^{++}. \tag{3}$$

The solution electron transfer shown in Eq. (1) should be ruled out because of the large difference in redox potentials of the reacting species ($\Delta Ep = 1.37$ V).

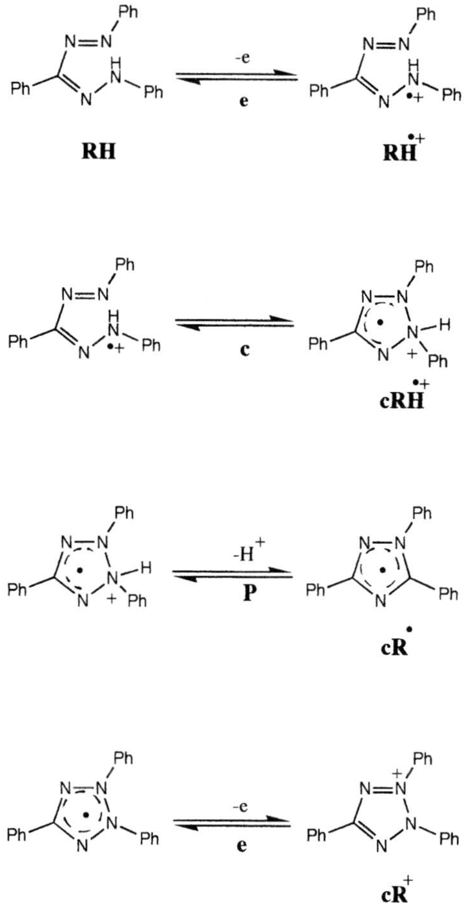

Scheme 3.

The disproportionation reaction (Eq. 2) of two tetrazolinyl radicals was studied by Umemoto [18a] and it was concluded that this reaction is a slow process, and therefore this process should also be ruled out as a fast **d**-step. Importantly, the difference between the redox potentials of cR$^+$ and R$^-$ (ΔEp = 0.25 V) favors the backward reaction. The feasibility of the backward reaction is substantiated by ESR experiments by Maender and Russell [20] who found that the mixture of formazan and tetrazolium salt gave rise to tetrazolinyl radicals. Finally, the solution electron transfer (Eq. 3) is possible as a homogeneous electron transfer (**d**-step) since it would be reasonable to expect that the redox potentials of the reacting species are very close. However, this reaction would imply dEp/dlog v slope of 19.7 mV (Table 1) which was not observed. Taking all the arguments into account it can be concluded that the mechanism shown in

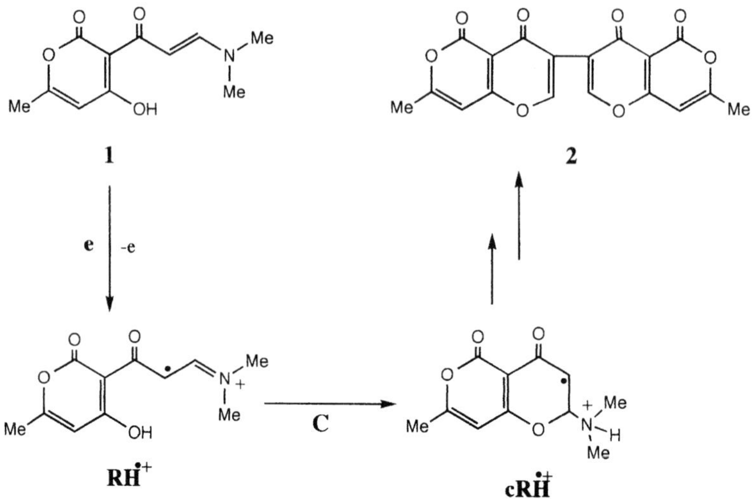

Scheme 4.

Scheme 3. is the most likely reaction pathway for the anodic cyclization of formazan to tetrazolium salt.

The electrode processes on the voltammetric and the preparative electrolysis time scales may be quite different. The oxidation of enaminone **1** with the hydroxy group in the ortho position under the controlled potential electrolysis gave bichromone **2** in 68% yield (Scheme 4.) with the consumption of 2.4 F/mol [21]. The RDE voltammogram of the solution of **1** in CH_3CN-0.1 mol/l Et_4ClO_4 showed one wave whose current function, $i_1/\omega^{1/2}C$, was constant with rotation rates in the range from 100 to 2700 rpm and showed one-electron behavior by comparison to the values of the current function with that obtained for ferrocene. The LSV analysis was undertaken in order to explain the mechanism of the reaction which involves several steps (**e-c**-dimerization-**p**-deamination). The variation of Ep/2 with log v was 30.1 ± 1.8 mV and variation of Ep/2 with log C was zero. Thus, our kinetic data obtained from LSV compare favorably with the theoretical value, 29.6 mV at 298 K, for a first order rate low [15]. This observation ruled out the dimerization of radical cation, **RH**$^{·+}$, for which theory predicts dEp/dlog v 19.7 mV and dEp/dlog C 19.7 mV variations. The most likely mechanism of the oxidation of **1** which fits the observed electroanalytical and preparative results, can be described in the form of Scheme 4.

Oxidation of enaminone **1** is initiated by electron loss from the dimethylamino moiety leading to radical cation, **RH**$^{·+}$. The following chemical reaction would be an intramolecular cyclization through addition of a hydroxy group on the radical cation site yielding a cyclic radical cation, **cRH**$^{·+}$. This step is most likely the rate-determining step. The cyclic radical cation then dimerizes to give an intermediate which subsequently undergoes an elimination of two protons and two molecules of dimethylamine, presumably in concerted fashion. In the slow preparative electrolysis, dimethylamine is gradually oxidized at the applied potential. Thus, as the chemical reaction takes place during exhaustive electrolysis, the apparent number of electrons determined by RDE voltammetry increases from limits n = 1 to n = 2.4 F/mol, determined by coulometry.

In the recent study on the anodic oxidation of enaminones which possess an unsaturated chain susceptible to react intramolecularly with an electrogenerated radical cation, the evidence for an intramolecular reaction was provided on the basis of the $dE_p/d\log v$ slope of 30 mV and one-electron behavior of the voltametric wave [48]. The reaction could involve similar mechanistic pathways as shown in Scheme 4 (**e-c**-dimerization and following chemical reactions). However, the authors were not able to isolate the products after preparative oxidation in order to confirm the possible mechanism.

2.2 Intermolecular Cyclizations

The intermolecular cyclization reactions could generally be characterized as two different types of process. The first would occur when the nucleophile present in solution attacks the anodically generated electrophile (radical cation, cation or neutral molecule), forming an intermediate adduct with proper geometry which then undergoes the ring closure reaction. The second type would involve the dimerization of the anodically generated radical cation with the ring closure as the completing chemical reaction. The mechanistic complexities involved in both types of process will be discussed in the following examples.

In 1981 we published the first paper [22] on the synthesis of s-triazolo[4,3-a]pyridinium salts, **4**, by the anodic oxidation of hydrazones **3** in the presence of pyridine (Scheme 5). In our working mechanistic scheme we proposed nitrilimine as the possible intermediate and pointed out that this reaction opens the door to a wide range of heterocyclic systems via anodic oxidation of aldehyde hydrazones through 1,3-dipolar cycloaddition reactions of the nitrilimine involved.

This paper prompted Jugelt, Grubert and co-workers to use the same reaction concept and accomplish several very good synthetic reactions [23–28]. When they performed t he oxidation of hydrazones in the presence of imidazole

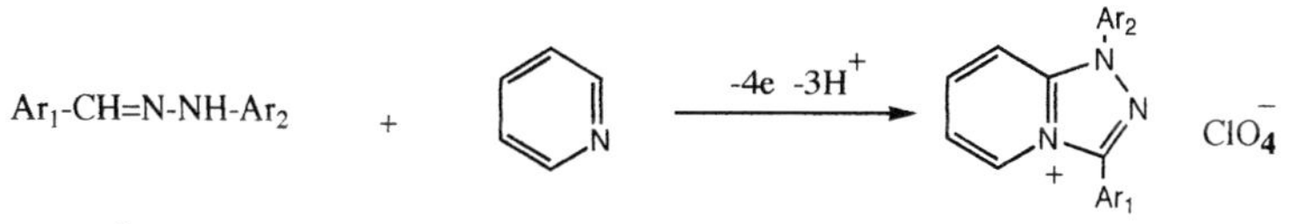

Ar_1-CH=N-NH-Ar_2 + [pyridine] $\xrightarrow{-4e\ -3H^+}$ [triazolopyridinium with Ar_2, Ar_1] ClO_4^-

3 **4**

Scheme 5.

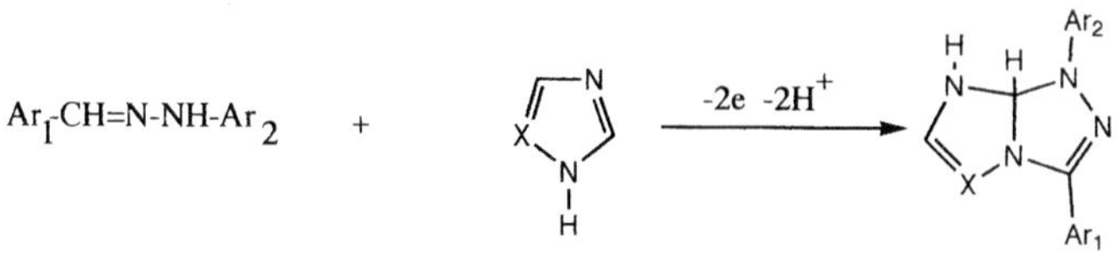

Ar₁-CH=N-NH-Ar₂   +   [imidazole/triazole structure, X=N or CH] $\xrightarrow{-2e\ -2H^+}$ [bicyclic product 5]

X=N or CH

Scheme 6.

and 1,2,4-triazole as dipolarophiles (or nucleophiles) the structure of the reported cycloadduct **5** as the reaction product [25] (Scheme 6) was not in agreement with the results obtained in our laboratory. We have therefore carried out the study of the anodic oxidation of hydrazones in the presence of heteroaromatic and Schiff bases with the purpose of examining these contradictory results [29].

The anodic oxidation of [p-(N,N-dimethylamino)benzylidene)-p-nitrophenylhydrazine, **3**, was studied in CH_3CN-0.1 mol/l Et_4ClO_4 solution in the presence of various heteroaromatic and Schiff bases. The cyclic voltammogram of **3** exibits two waves at 0.59 and 1.27 V vs SCE. The first peak corresponds to the oxidation of the parent molecule and the second to **3** protonated by protons liberated along the first wave. Sweep reversal from the anodic to the cathodic side causes the appearance of one reduction peak at 0.43 V vs SCE. The current function, $i_1/\omega^{1/2}C$, showed a one-electron behavior (n = 1). After addition of a heteroaromatic base (pyridine, imidazole or 1,2,4-triazole) the first wave doubled in height (n = 2) and the reduction peak at 0.43 V disappeared. After addition of the Schiff base the height of the first wave remained practically the same but the peak at 0.43 V was reduced. With regards to the reaction mechanism for the formation of compound **4**, the following questions seemed to interest – is the oxidation of **3a** two-electron and one-proton loss leading to 1-aza-azoniaallene cation, **3a**, as an intermediate and is the oxidation of **3** a two-electron and two-proton loss leading to nitrilimine, **3b**, as an intermediate (Scheme 7)?

1-Aza-2-azoniaallene cations, **3a**, are generated as reactive intermediates in many of the oxidative processes of hydrazones [30–33]. The cations **3a** react with acetylene, olefins, isocyanates, carbodiimides and nitriles under [3 + 2]-cycloadditions [34]. The cycloaddition reaction is stereoselective and according to AM1 calculations, and the cycloaddition of acetylenes to 1-aza-2-azoniaallene cations is a concerted process, which can be classified as a "1,3-dipolar cycloaddition with reverse electron demand" [35]. The formation of the stereoselective cycloaddition products obtained in the anodic oxidation of hydrazones in the presence of norbornene [24] could be interpreted to occur through generation of a cation **3a** without involving the nitrilimine, **3b**, as an intermediate. The 1,3-dipolar cycloaddition reactions of nitrilimines are generally thermally allowed pericyclic reactions [36] and they can be controlled either by HOMO

Anodic Synthesis of Heterocyclic Compounds

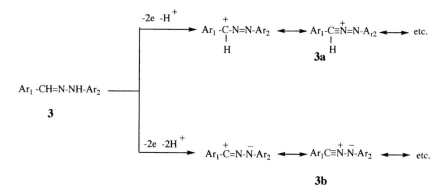

Ar_1=p-(CH_3)$_2$NC_6H_4

Ar_2=p-$NO_2C_6H_4$

Scheme 7.

(dipole)-LUMO (dipolarophile) or HOMO(dipolarophile)-LUMO(dipole) interaction [37]. The dominant interaction is the one that involves orbitals much closer in energy and a maximal HOMO-LUMO overlap.

We believe that the intermediate generated by anodic oxidation of **3** is the cation **3a** and not nitrilimine **3b**. The generation of the cation intermediate was postulated in several papers on the anodic oxidation of hydrazone derivatives [38–41]. W also believe that both heteroaromatic and Schiff bases react with the cation **3a** as nucleophiles rather than dipolarophiles. The most likely mechanism of the oxidation of the hydrazone **3** in the presence of heteroaromatic and Schiff bases, which would best fit the observed electrochemical and preparative results, can be described in the form of Scheme 8.

Hydrazone **3** is oxidized at the anode through the loss of two electrons and one proton giving rise to cation **3a**, which can then be detected as a stable species by slow cyclic voltammetry as a reduction peak at 0.43 V vs SCE. The stability of **3** is enhanced by the resonance contribution of the lone pair of the electrons on the p-N,N-dimethylaminophenyl ring. Pyridine attacks the cation **3a** leading to the new cation **3c**. It seems that pyridine reacts faster as a nucleophile than as a base, which would then result in a nitrilimine as an intermediate. The cation **3c**, as an acidic species, deprotonates leading to the intermediate **3d**, which may cyclize through a 1,5-dipolar addition to the compound **3e**. It has been shown [26] that the oxidation potential of compound **3e** is lower than the applied potential, so that **3e** is further oxidized through a two-electron and one-proton loss, leading to the final product **4**. In the presence of 1,2,4-triazole, cation **3a** is attacked leading through deprotonation to the intermediate **3f**, presumably as a kinetic product, which irreversibly tautomerizes to more stable hydrazone **5** as a final product. In our original paper [29] we assigned the structure **3f** to the

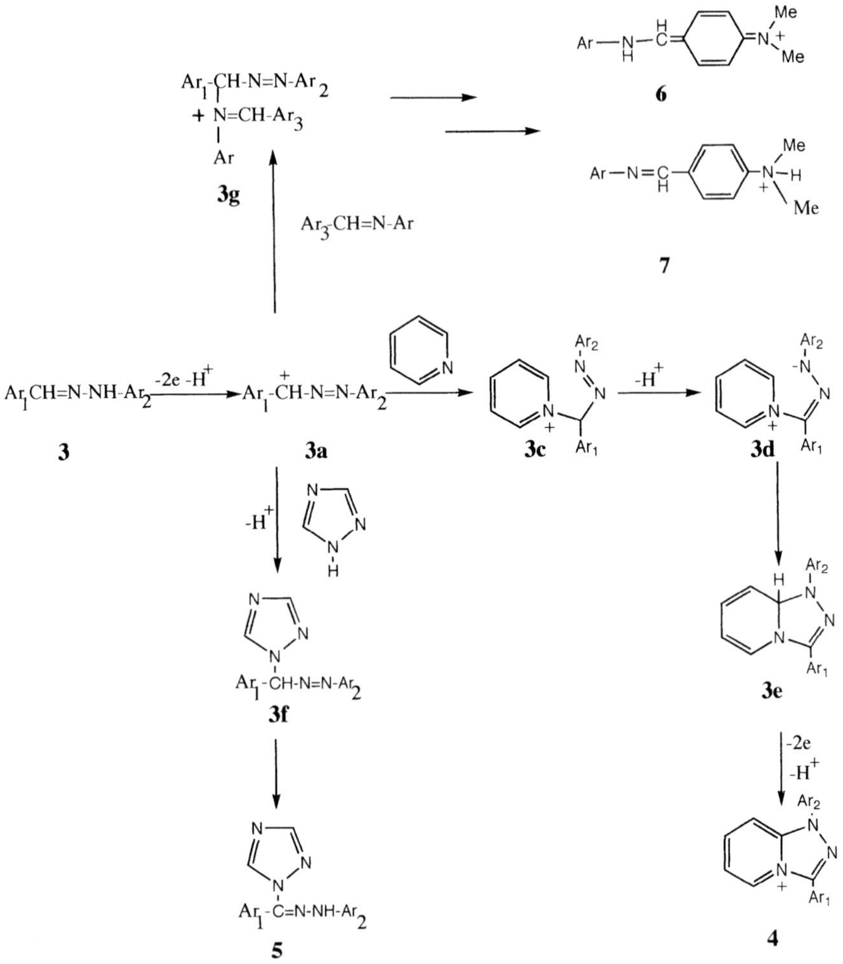

Ibro Tabaković

Scheme 8.

final product. However, we later revised the structure of the final product, which is actually **5**, on the basis of X-ray analysis [42]. Similar products were also obtained by reaction on N-benzoylbenzhydrazidoyl chloride with imidazole, 1,2,4-triazole, and benzotriazole in the presence of triethylamine [43]. These reaction conditions are typical for the formation of nitrilimine as an intermediate. However, as the authors pointed out, the concomitant reaction of the azole and hydrazidoyl chloride cannot be discarded. The isolated products **6** and **7**, after oxidation of hydrazone **3** in the presence of Schiff bases, indicate that the intermediate cation **3a** is the kinetic species.

Nitrilimine **3b** was the most probable intermediate when the hydrazone **5** was oxidized using controlled potential electrolysis. The one-electron oxida-

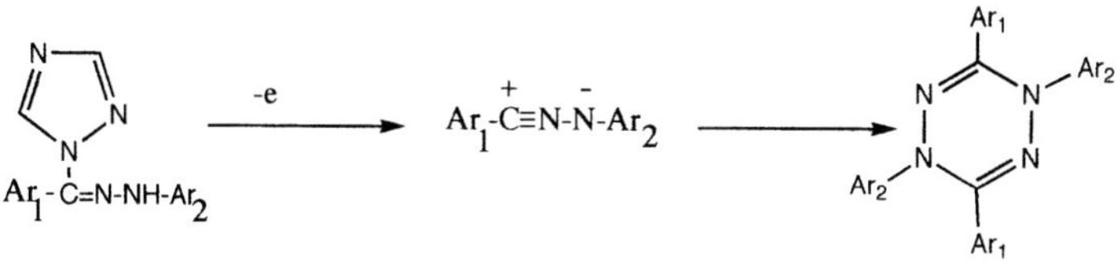

Anodic Synthesis of Heterocyclic Compounds

Scheme 9.

tion of **5** (n = 1.1 F/mol), followed presumably by loss of one proton and the 1,2,4-triazolyl radical, gave rise to nitrilimine **3b** which then dimerizes to 1,4-dihydro-1,2,4,5-tetrazine derivative **8** in 92% yield (Scheme 9). The same types of the tetrazine products were obtained by anodic oxidation of N-aroylbenzhydrazidoyl chloride [43b]. Importantly, the formation of the product **8** was not observed during the oxidation of hydrazone **3** in the presence of heteroaromatic bases.

Another example of an intermolecular cyclization is the anodic generation of a neutral molecule acting as an electrophile which reacts with the nucleophile present. We have shown [44] that the anodic oxidation of cathecol (**QH$_2$**) at a graphite anode using controlled potential electrolysis in the presence of 4-hydroxycoumarin gave 6H-benzofuro[3, 2c][1]-benzopyran-6-one, **9**, in yields higher than 90%, irrespective of the applied potential (E = 0.4 or 1.1 V vs SCE) with consumption of 4 F/mol at either potentials (Scheme 10).

In order to explain the formation of the product **9** during oxidation at two different potentials we have performed the experiments with cyclic voltammetry [45]. The cyclic voltammogram of catechol (**QH$_2$**) exhibits the anodic wave at 0.25 V vs SCE (Fig. 1a) corresponding to the formation of o-quinone (**Q**) which is reduced in the cathodic sweep at 0.05 V vs SCE. The cathodic counterpart of the anodic peak disappears, when a sufficient amount of 4-hydroxycoumarin was added, and a second irreversible peak at 0.95 V vs SCE appeared (Fig. 1b).

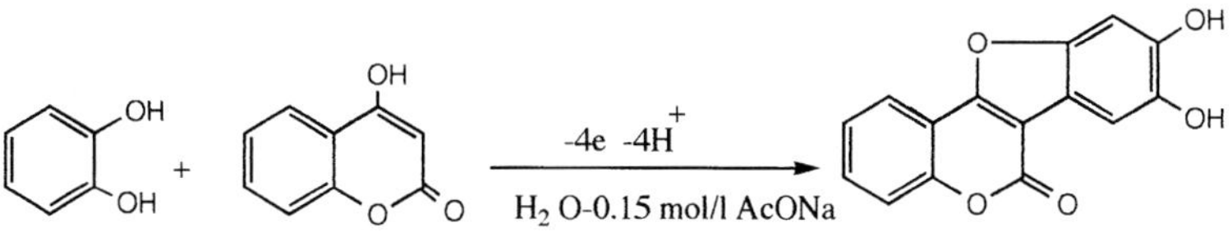

E=0.4 V vs.SCE (90%)

E=1.1 V vs. SCE (95%)

Scheme 10.

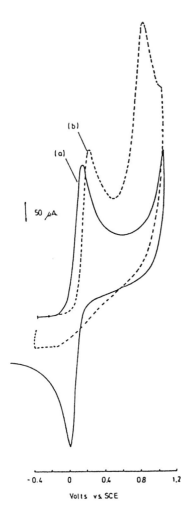

Fig. 1. Cyclic voltammogram of 1 mmol/l catechol in 0.15 mol/l aqueous sodium acetate solution at pyrolitic graphite anode (*curve a*) and plus 1 equivalent of 4-hydroxycoumarin (*curve b*). Scan rate: 150 mV/s [45]

The combination of the arguments obtained from preparative electrolysis, coulometry at controlled potential, and cyclic voltammetry allowed us to formulate the main features of the reaction pathways as in Scheme 11. The results indicated that at 0.4 and 1.1 V, two distinct oxidative processes were occurring, both of which resulted in the same product **9**. The anodic oxidation of catechol, **QH$_2$**, at 0.4 V vs SCE leads to the corresponding *o*-quinone, **Q**, which acts as a Michael acceptor with 4-hydroxycoumarin giving rise to the intermediate adduct **9a**. However, the structure of adduct **9a** is not quite certain. In our original paper [45] we envisaged the structure of the adduct as a product of double Michael addition. However, it is believed that the adduct **9a** formed upon tautomerization (Scheme 11) is the more probable intermediate which is being oxidized at higher potential (E = 1.1 V vs SCE), due to the electron-

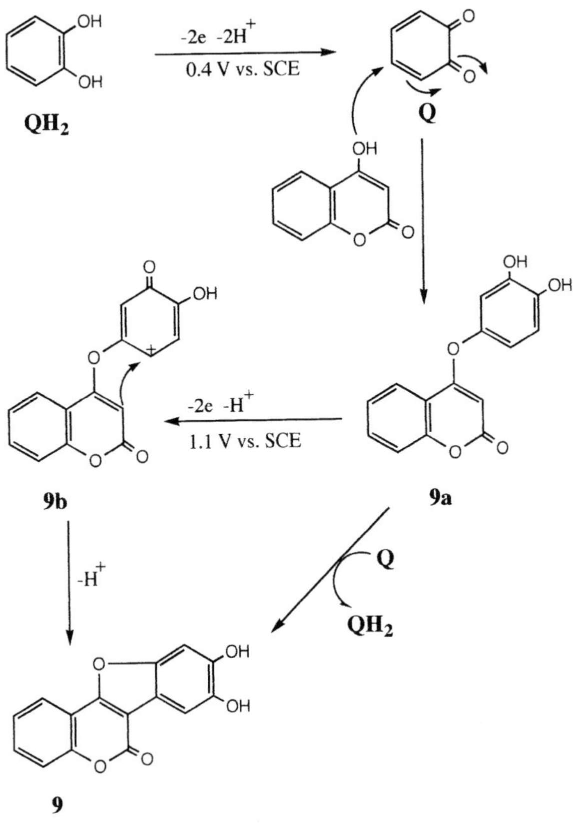

Scheme 11.

withdrawing property of the coumarin ring, giving rise to the intermediate cation **9b** which cyclizes to the final product **9**. We also proposed that the intermediate **9a** could be oxidized through homogeneous oxidation in which **QH$_2$** is regenerated, and hence can be reoxidized at the lower applied potential (E = 0.4 V).

The radical cations generated at the anode surface can dimerize before they diffuse into solution. The dimeric compound formed can then be cyclized in solution via formation of a carbon-heteroatom bond. The anodic oxidation of enaminones reported earlier [46] and also recently [47] could serve as an illustration of this type of intermolecular cyclization.

Anodic oxidation of enaminone **10** was performed at the platinum gauze electrode in methanol solution containing LiClO$_4$ in a divided cell at controlled potential. After passage of 1.2 F/mol two major products **11** and **12** were isolated in 50 and 20% yields respectively (Scheme 12). Oxidation of the enaminone **10** is initiated by electron loss from the dimethylamino moiety to

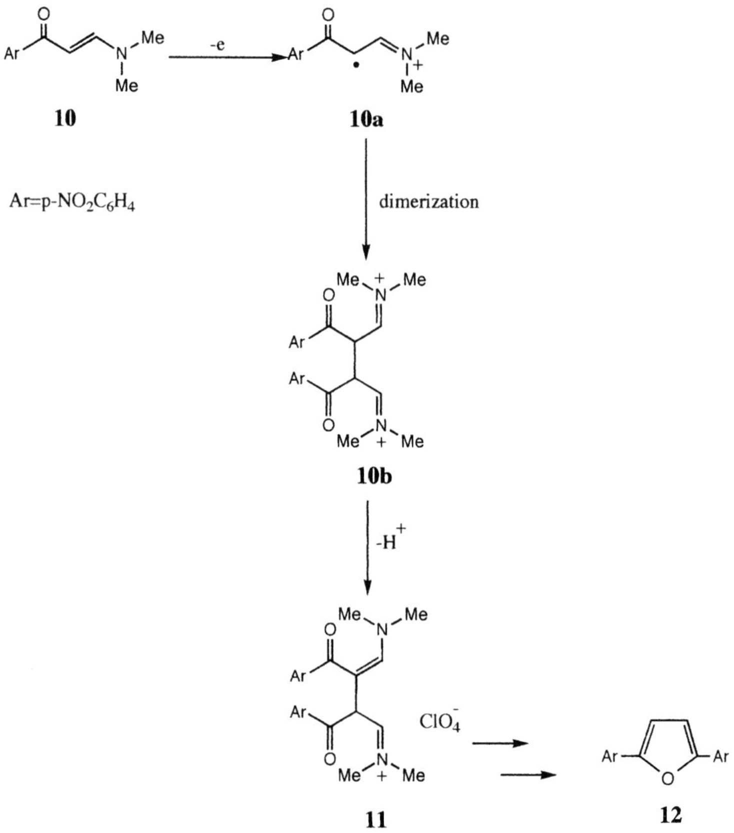

Ibro Tabaković

Scheme 12.

Ar=p-NO$_2$C$_6$H$_4$

give the radical cation **10a** which is presumably the rate-determining step of the overall reaction ($\alpha n = 0.64$, $n = 1$). The resulting radical cation dimerizes to the dication **10b** which is deprotonated leading to the product **11**. The formation of bis-(2,5-p-nitrophenyl)furan **12** is most likely the result of a multistep base-promoted transformation of **11** to **12**.

2.3 Heterogeneous vs Homogeneous Oxidation

In the following section, the chemistry of the generated radical cations which cyclize will be analyzed using the selected examples in which the radical cations are formed through heterogeneous anodic oxidation ("direct electrolysis"), or through homogeneous oxidation by means of well-defined outer-sphere one-electron oxidants added either in stoichiometric amounts or generated in situ by "indirect electrolysis" [49]. As schematically demonstrated in Fig. 2, the indirect

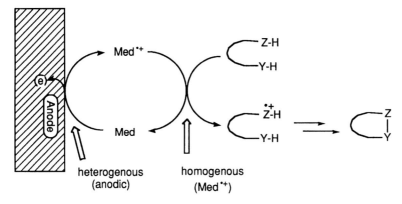

Fig. 2. Heterogenous (anodic) and homogeneous (Med$^{\cdot+}$) electron transfer steps in a mediated oxidation

electrolysis combines at heterogeneous step, that is the anodic oxidation and regeneration of the mediator at such a low potential that the substrate is electrochemically inactive, with the homogeneous redox reaction of the substrate with the generated oxidant (**Med$^{\cdot+}$**). Since both, anode and **Med$^{\cdot+}$** are essentially outer-spheres oxidants, the major difference in the product formation would be the consequence of the different distribution of the generated substrate radical cations. Thus, the radical cations are formed at rather high concentration on the anode surface, while in the homogeneous reaction their distribution is uniform in solution.

Anodic oxidation of chalcone phenylhydrazone **13** on the voltammetry time scale corresponds to the formation of pyrazole **14** through an **e-c-p-D-p** mechanism [50]. However, in the preparative controlled potential electrolysis, the protons liberated during the anodic oxidation catalyze cyclization of **13** to form 1,3,5-triphenyl-Δ2-pyrazoline **15**, which is oxidized to the dimer **16**. The compound **16** showed two reversible waves with the peak potentials at 0.45 and 0.65 V vs SCE. Since the applied potential in preparative electrolysis was 0.95 V vs SCE the compound **16** was further oxidized to give **17**. Anodic oxidation at a platinum electrode in CH_3CN-0.1 mol/l Et_4ClO_4 solution gave **14**, **16**, and **17** in 15, 30, and 35% yields, respectively. In the presence of pyridine, pyrazole **14** was obtained in 40% yield. The oxidation of **15** gave only dimeric products **16** and **17** in 32 and 40% yield, respectively (Scheme 13). The spin density in **15**$^{\cdot+}$ appears to be concentrated in the N-phenyl ring and in the absence of base the dimerization at the para position of the ring is the major reaction pathway. We have shown that 1,5-diphenyl-3-(4-hydroxycoumarinyl)-pyrazoline can be anodically oxidized in the presence of pyridine to give the corresponding pyrazole in 95% yield [51].

The homogeneous oxidation of compounds **14** and **15** with thianthrene radical cation perchlorate (**Th$^{\cdot+}$ClO$_4^-$**) was studied later by Kovelski and Shine

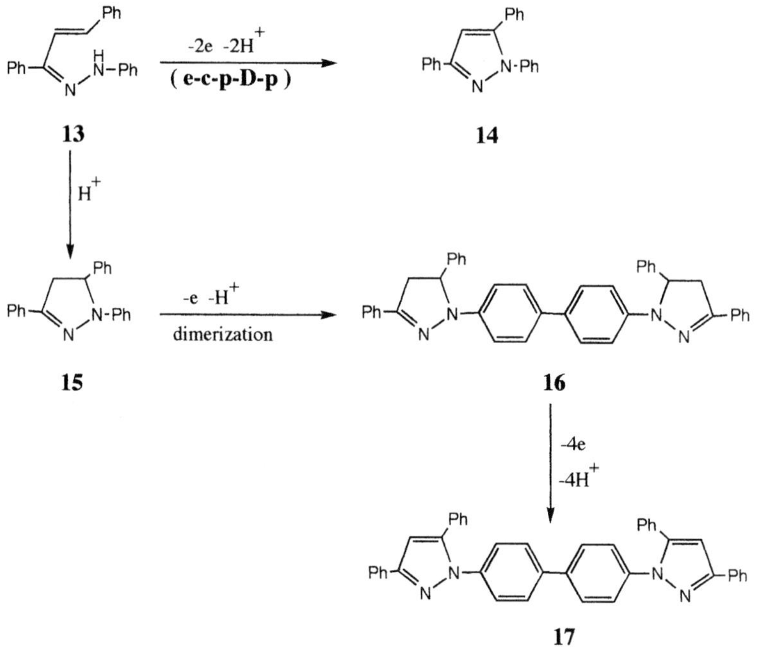

Scheme 13.

[52] in both CH_3CN and CH_2Cl_2 solutions. Oxidation of chalcone phenylhydrazone **13** with $Th^{·+}$ in the reactant ratio $Th^{·+}/\mathbf{13}$ from 2.0 to 3.0 gave the pyrazole **14** in excellent yields ranging from 66 to 98%. It seems that the rate of the second electron transfer, which appeared to be the rate-determining step (D-step) in the anodic oxidation, is increased by homogeneous oxidation with $Th^{·+}$ present which resulted in high yield of **14**. The oxidation of the pyrazoline **15** with $Th^{·+}$ also gave an excellent yield of **16**. This result suggests that the lower concentration of the initially formed radical cations $\mathbf{15}^{·+}$ in homogeneous solution, compared to heterogeneous anodic oxidation, is not the critical factor for the dimerization reaction. The relative stability of the radical cation $\mathbf{15}^{·+}$ is presumably the governing factor for the dimerization reaction. Shin and Shine also studied the oxidation of the trisubstituted phenols having *tert*-butyl groups in 2- and 6-positions with $Th^{·+}$, which were converted into a benzoxazole by reaction with solvent nitrile and oxidative loss of *tert*-butyl group [53]. This reaction is similar to the previously described anodic oxidations [54].

Anodic oxidation of formazane **18** [17], 1-arylmethylenesemicarbazide **19** [55], *p*-nitrobenzaldehyde phenylhydrazone **20** [56], and 2-benzoylpyridine phenylhydrazone **21** [57] afforded the corresponding heterocycles in a very good yield (Scheme 14). The homogeneous oxidation of compounds **18–20** was carried out by indirect electrolysis by the mediators generated in situ [58].

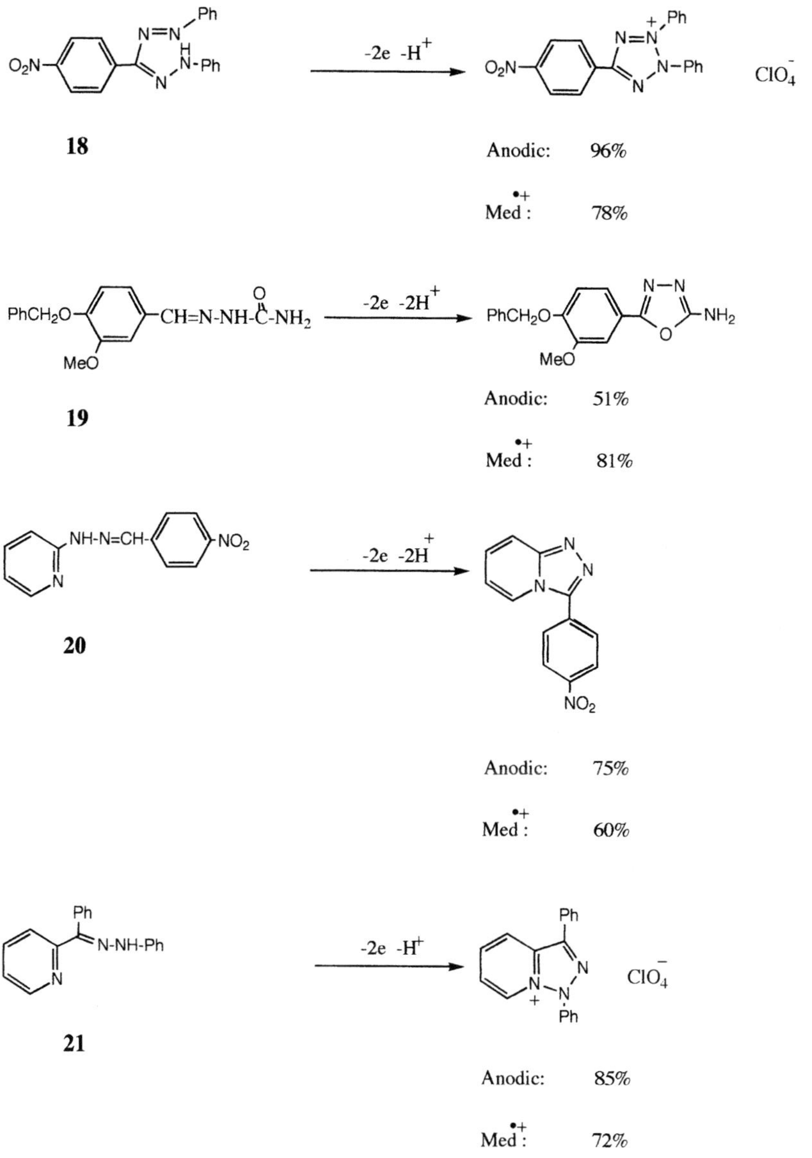

Anodic Synthesis of Heterocyclic Compounds

Scheme 14.

Specifically, the mediators (**Med**$^{•+}$) used were the radical cations of tris-(4-bromophenyl)amine and 2,3-dihydro-2,2-dimethylphenothiazine-6(1H)-one [59]. The results of the oxidative cyclizations under the homogeneous oxidation conditions are parallel to those obtained by direct anodic oxidation.

Some years ago we studied the homogeneous oxidation of 2'-hydroxychalcones with tris-(p-bromophenyl)amine radical cation (**Med**$^{\bullet+}$) generated in situ in CH_3OH–CH_2Cl_2 (3:1)-0.2 mol/l $LiClO_4$ solution [60]. We considered these experimental conditions as a model for biomimetic transformations with the purpose of studying the equilibrium between 2'-hydroxychalcone and the isomeric flavanone and to answer the question of whether the flavanones are true intermediates in the biosynthesis of flavanoids. The direct anodic oxidation of 2'-hydroxychalcone was also carried out [61] and the results shown in Scheme 15 illustrate the differences between homogeneous and heterogeneous oxidations.

Anodic oxidation of chalcone **22** gave flavone **24**, but flavanone **23** was not detected. The results obtained through the homogeneous oxidation using **Med**$^{\bullet+}$ showed that the oxidation of **22** to **24** occurred through flavanone **23** as an intermediate. We have shown that flavanones can be oxidized with **Med**$^{\bullet+}$ generated in situ to flavones and flavanols in high yield [62]. The remaining question is whether or not the transformation of 2'-hydroxychalcone **22** to the

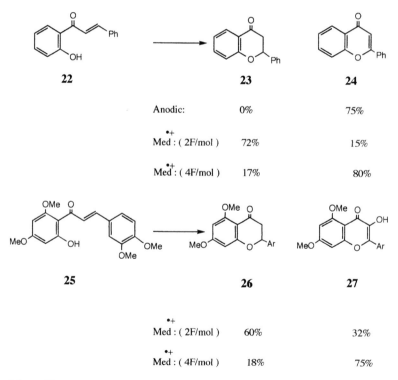

Scheme 15.

isomeric flavanone **23** is the result of an acid-catalyzed cyclization due to protons liberated during the oxidation, or the result of an electron transfer reaction. The results may be rationalized by considering the fate of the initially formed radical cation **22a** (Scheme 16).

The reversible wave of **Med** appears at 1.1 V vs SCE. Addition of chalcone **22** leads to the enhancement of the oxidation peak while its reverse cathodic peak is suppressed. It is evident that **Med**$^{\cdot+}$ oxidizes the chalcone **22**, whose direct oxidation occurs at a higher potential (Ep = 1.7 V vs SCE) to the radical cation **22a**, which subsequently undergoes an intramolecular cyclization leading to radical **22b**, which then abstracts the hydrogen from the solvent giving rise to flavanone **23**. By means of GLC-MS analysis of the reaction mixture after indirect electrolysis in deuterated solvents (CD_3OH–CD_2Cl_2) the flavanone with an incorporated deuterium atom was detected [63]. Our findings were substantiated by results of photoinduced single-electron transfer transformations of 2′-hydroxychalcones to flavanones [64]. It seems that the distribution of the generated radicals in homogeneous and heterogeneous oxidation of **22** is the important factor which has influence on the product distribution. The radical **22b** under homogeneous oxidation conditions then undergoes the hydrogen abstraction reaction, while **22b** during anodic oxidation is further oxidized at the applied potential.

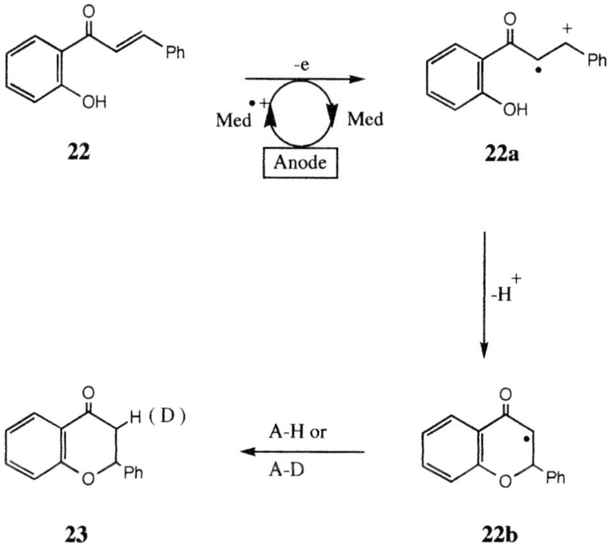

Scheme 16.

3 Anodic Synthesis of Heterocycles

Section 3 contains a discussion of the anodic reactions in which a heterocyclic system is formed. These reactions may be classified in different ways; here they are treated according to the type of bond formed.

3.1 Formation of the Carbon–Nitrogen Bond

3.1.1 Oxidation of Amines and Hydroxyamines

Nitrogen-centered radical reactions are less well developed than those of their carbon relatives even though the preparative potential for nitrogen radical transformations is high in the areas of the alkaloid and heterocyclic chemistry of the pyrrolidine nucleus [65]. Suginome and coworkers have shown that aminyl radicals can be generated by anodic oxidation of the lithium amides **29a** of δ-alkenylamines **29** [66–68] (Scheme 17). The constant current anodic oxidation of **29** in THF-HMPA (30:1)-0.25 mol/l LiClO$_4$ solution gave 2,5-disubstituted pyrrolidines in 6–85% yield. Yields were generally better with 2-phenyl or 2-methyl alkenylamines, and with a phenyl-substituted double bond. The process is highly regioselective and stereoselective in forming cis-2,5-disubstituted pyrrolidines from 5-exo cyclization of **29b** radicals. Piperidine products, as a result of the possible 6-endo cyclizations, were not observed. Notably, an attempted cyclization by direct anodic oxidation of δ-alkenylamines **29** failed to give any cyclization product [68].

The cyclization of the lithium amide **29a** to 2,5-disubstituted pyrrolidine shown in Scheme 17 is clearly a one-electron oxidation process. This suggests that the radical **29c** is not an oxidizable species at the applied potential and thus

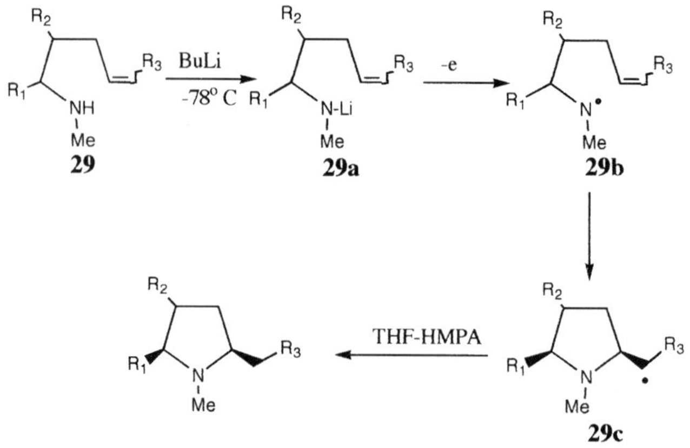

Scheme 17.

diffuses from the electrode into solution leading through H-atom abstraction to the final product. Another explanation for the one-electron oxidation process might be that the initially generated nucleophilic aminyl radical **29b** is complexed to the lithium cation, as a Lewis acid, giving an electrophilic species with presumably a more positive oxidation potential than the substrate **29a**.

The anodic oxidation of N-methoxy derivatives of δ-alkenylamines gave rise to a single stereoisomer of 5-substituted-1-methoxypyrrolidines in 42% (R = CH$_3$) and 38% yield (R = Ph), respectively [68] (Scheme 18).

Anodic oxidation of a suitably unsaturated hydroxyamine was designed for the synthesis of the racemate of an 11-hydroxy metabolite of the important N-methyl-D-aspartate receptor antagonist MK-0801 [69]. Anodic oxidation of

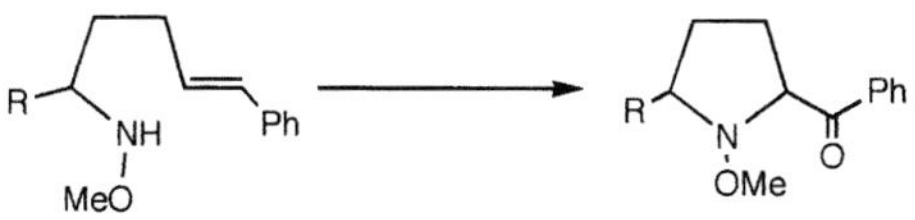

Scheme 18.

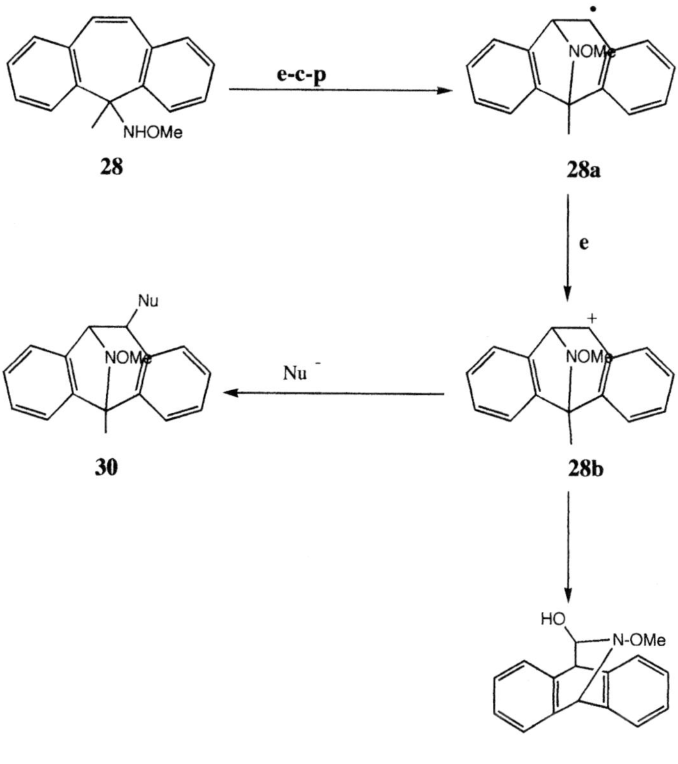

Scheme 19.

methoxylamine **28** in aqueous THF afforded the desired derivative **30** as a major product and the by-product **31** (Scheme 19).

The authors proposed mechanistic pathways for the cyclization reaction which involves direct electron transfer from the methoxylamino moiety (**e**-step), cyclization of the generated radical cation (**c**-step), and deprotonation (**p**-step) leading to the radical **28a**. The electron-rich carbon radical is readily oxidized to carbocation **28b**, which is then trapped with water yielding the major product **30**. The by-product **31** is clearly the result of a rearrangement of the intermediate carbocation **28b**. This result presents a direct proof that the cyclization reaction occurs after the first electron transfer. However, the question whether the reactive species in the cyclization step is a radical cation or an aminyl radical, i.e., discrimination between **e-c-p** and **e-p-c** mechanisms, remains open at the present time. Optimization using a flow cell design and conditioned graphite felt gave **30** in 85% yield on a 200 g scale [70].

3.1.2 Oxidation of Hydrazones and Schiff Bases

Several heterocyclic hydrazones **32** have been oxidized in CH_3CN-0.1 mol/l Et_4ClO_4 solution with addition of 60% $HClO_4$ to s-triazolo[4,3-a]pyridine derivatives **33** in yields ranging from 55 to 92% [56] (Scheme 20).

The best yields of the products were obtained in acidic media in which electrolysis proceeded smoothly without inhibition of the electrode. The electrolysis without $HClO_4$ present usually resulted in a lower yield of the product **33** and filming of the electrode occurred during the reaction. The parent compound is the protonated hydrazone **32** and the overall reaction includes the loss of two electrons, three protons, and one cyclization step. Based on the results obtained from LSV and CPSV it was concluded that oxidative cyclization is taking place after the second electron transfer via the **e-p-D-p-c-p** mechanistic scheme. Four-electron oxidation of hydrazones in the presence of pyridine, substituted pyridines, quinoline, isoquinoline, and benzo[f]quinoline gave, through intermolecular cyclization, several s-triazolo[4,3-a]pyridinium perchlorates, s-triazolo[4,3-a]quinolinium perchlorate, s-triazolo[3,4-a]isoquinolinium perchlorate and s-triazolo[4,3-a]benzo[f]quinolinium perchlorate in yields ranging from 30 to 90% [26–29]. Anodic oxidation of the sodium salt of tosylhydrazone of tropone **34** and 2-phenyltropone **36** afforded 2-tosyl-2H-

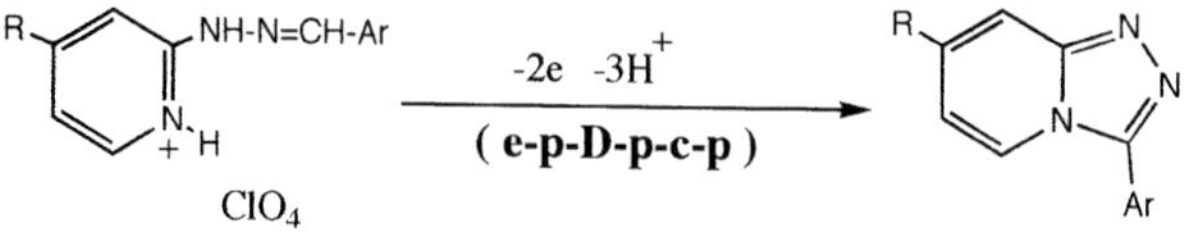

Scheme 20.

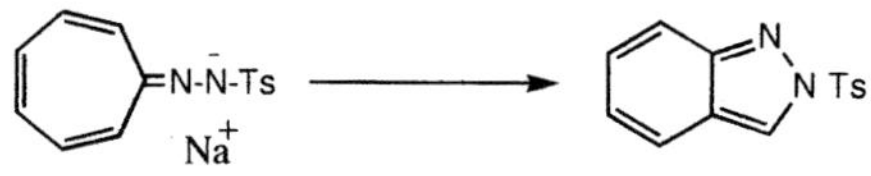

Scheme 21.

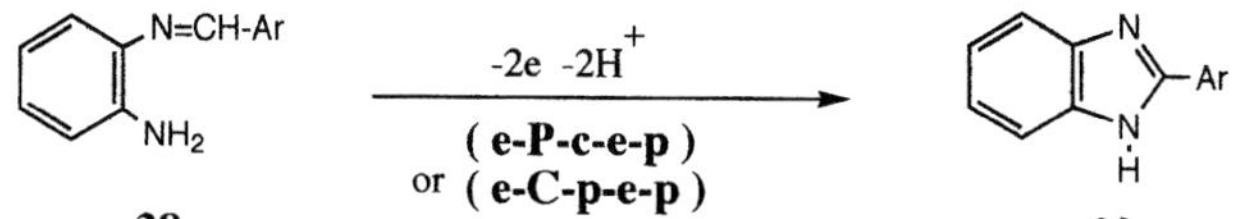

Scheme 22.

indazole **35** and 1-tosyl-7-phenyl-1*H*-indazole **37** in 28 and 17% yield, respectively [71] (Scheme 21).

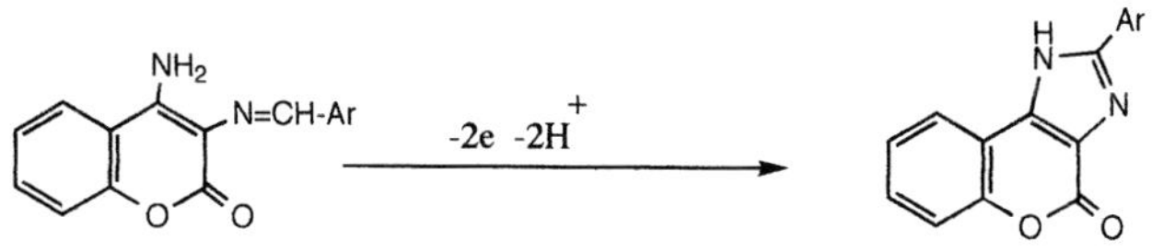

Anodic oxidation of *o*-amino substituted aromatic Schiff bases (**38** and **40**) to imidazole derivatives **39** and **41** were carried out in CH_3CN-0.1 mol/l Et_4ClO_4 solution with addition of pyridine as a base, using controlled potential electrolysis and a divided cell [72] (Scheme 22).

Oxidation in the presence of pyridine gave the products in 60–85% yield, whereas the electrolysis without pyridine lowered the yield to 10–20% and the products of hydrolysis, because of the accumulation of the acid in the anodic compartment, were identified. The reaction mechanism was proposed on the basis of LSV and CPSV results. The values of $dEp/d\log v = 30$ mV and $dEp/d\log C = 0$ mV point to the occurrence of a first-order rate-determining step. Comparison of the CPSV slope values of 58 mV with the theoretical value

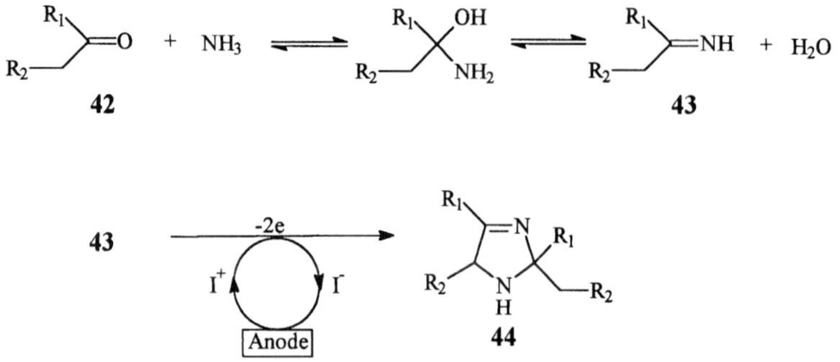

Scheme 23.

(58.6 mV) confirmed the conclusion of a first-order rate-determining step. However, because of the experimental difficulties involved, we were not able to discriminate between the **e-P-c-e-p** and the **e-C-p-e-p** mechanism, in order to answer the question of whether cyclization or deprotonation of the radical cation is the rate-determining step.

The indirect anodic oxidation of ketones **42** in ammonia – containing methanol using iodide as a mediator afforded 2,5-dihydro-1H-imidazols **44** via oxidation of the intermediate ketimine **43** to N-iodo imine followed by elimination of HI to afford the nitrenium ion, which subsequently reacts with ketimine **43** to give the product **44** [73] (Scheme 23).

The electrolysis of asymmetric ketones **43** led to the formation of isomers and stereoisomers. Kinetic measurements for the formation of ketimine **43** in saturated ammoniacal methanol indicated that at least 12 h of the reaction time were required to reach the equilibrium in which approximately 40% of **42** was converted into the ketimine **43**. However, the electrolysis was completed within 2.5 h and the products **44** were isolated in 50–76% yields. It seems that the sluggish equilibrium gives a significant concentration of ketimine **43** which is oxidized by the I$^+$ generated at the anode, and the equilibrium is shifted towards formation of the product **44**. 2,5-Dihydro-1H-imidazols of type **44**, which were unsubstituted on nitrogen, are rare compounds. They can be hydrolyzed with hydrochloric acid to afford the corresponding α-amino ketones as versatile synthetic intermediates for a wide variety of heterocyclic compounds, that are otherwise difficult to prepare.

3.1.3 Oxidation of Imidamides

Anodic oxidation of N,N-disubstituted trifluoroethanimidamide **45** in dry and in aqueous acetonitrile gave the imidazole **46** and quinoneimine **47** as the reaction products (Scheme 24). The constant current electrolysis on a glassy carbon anode and a platinum cathode was performed in an undivided cell [74].

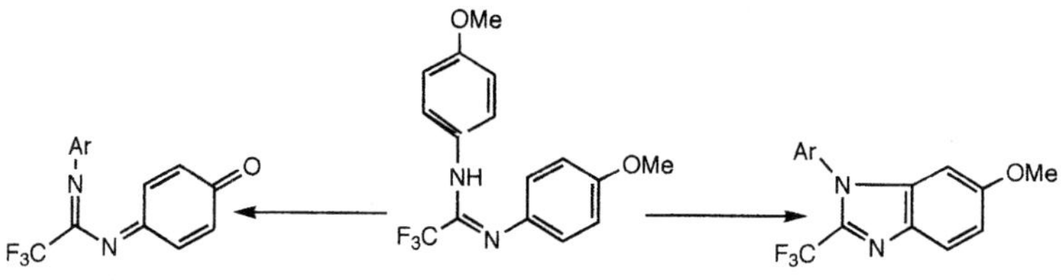

Scheme 24.

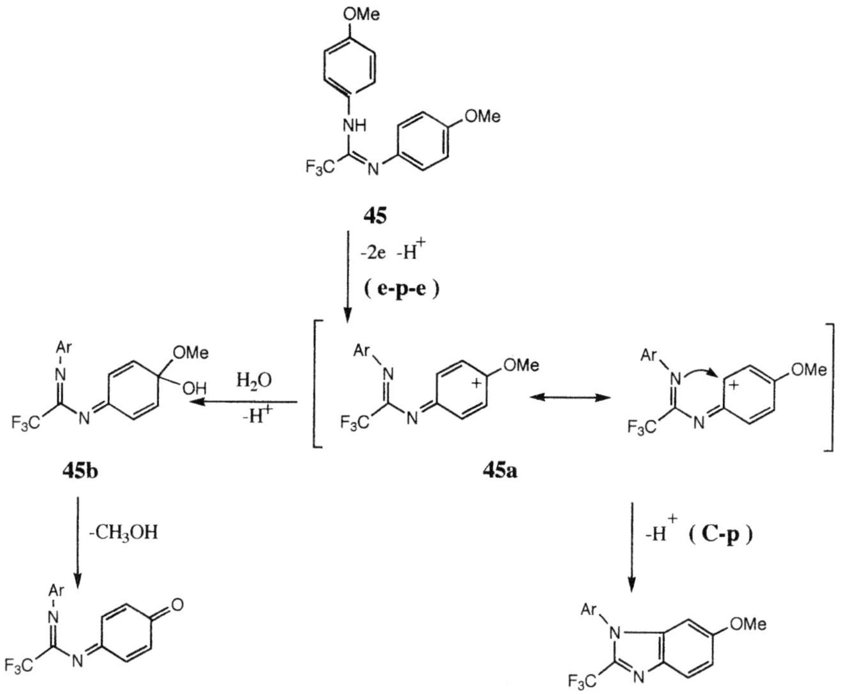

Scheme 25.

Anodic oxidation of **45** in dry acetonitrile at 60 °C and at low current density provided a quantitative yield of **46**, while oxidation of **45** in aqueous acetonitrile at 0 °C provided a high yield of **47**. It has been shown that quinoneimine **47** can be transformed to **46** in 93% yield, through BF_3Et_2O catalyzed cyclization [75]. The reaction pathways leading to the formation of **46** or **47** are summarized in Scheme 25. Two-electron oxidation of **45** leads to the cation **45a** through an ECE or **e-p-e** mechanism. It seems that the cyclization of **45a** is the rate-determining step in the overall intramolecular cyclization of **45** to **46**. The high reaction temperature (60 °C) required to perform the reaction suggests a high activation energy of the cyclization step. The nucleophilic attack of water on the cation **45a** seems to be much faster than the cyclization reaction leading to the intermediate, which is transformed to the final product **47** through elimination of methanol. Importantly, the best yields of **47** were obtained at low temperature (0 °C) for the electrolysis.

3.1.4 Oxidation of Hetero-Allenes

The anodic oxidation of substituted allenes [76–79] and hetero-allenes [80–85] has been extensively studied by Becker and coworkers. Oxidation of secondary and tertiary alkyl isothiocyanates resulted in an α-cleavage processes [77]. Primary alkyl isocyanates yielded amides and carbonyl compounds due to the nucleophilic involvement of either acetonitrile or water [80]. Primary alkyl isothiocyanates **48** afforded five-membered thiadiazolidine **49** and **50** and

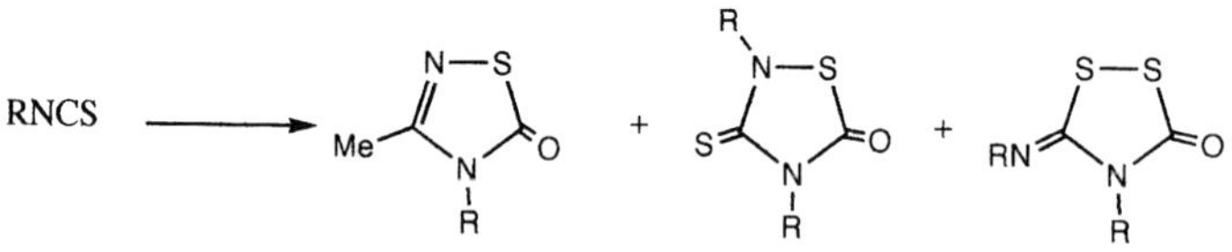

Scheme 26.

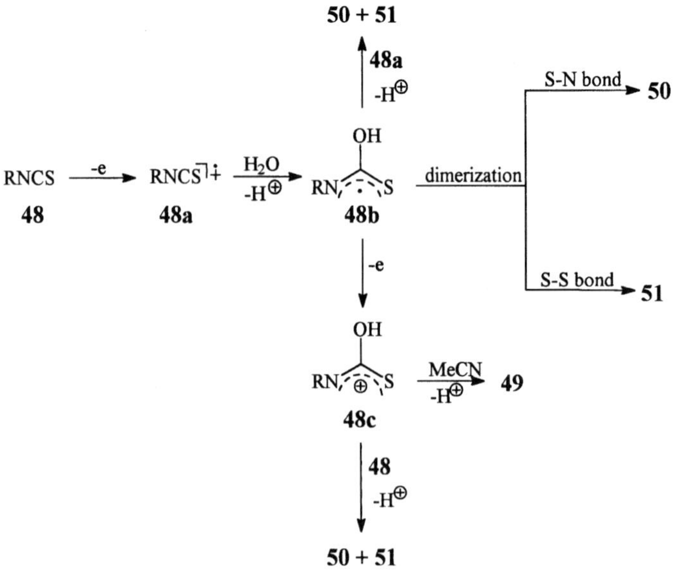

Scheme 27.

dithiazolidine **51** derivatives (Scheme 26). Anodic oxidation in acetonitrile solvent gave the compound **49** which contains one molecule of substrate, one molecule of solvent, and two additional products [80]. Anodic oxidation of RNCS in dichloromethane yielded two isomers **50** and **51** in 11–86 ad 28–90% yields, respectively, depending on the length or bulkiness of the alkyl groups [81].

The mechanism which could explain the formation of these products is described in Scheme 27. In an EC mechanism, the intermediate radical cation **48a** could undergo a follow-up reaction with water as a nucleophile to form radical **48b** which could than dimerize through S–N or S–S bond formation or react with **48a** to yield **50** and **51** as the fianl one-electron oxidation products. In an ECE mechanism, intermediate **48b** is further oxidized to **48c** which reacts with acetonitrile as a solvent to give **49** as the final two-electron oxidation product. The cation intermediate **48c** can react with the parent molecule **48** through [2 + 3]-cycloaddition to give the final products **50** and **51**. The [2 + 3]-

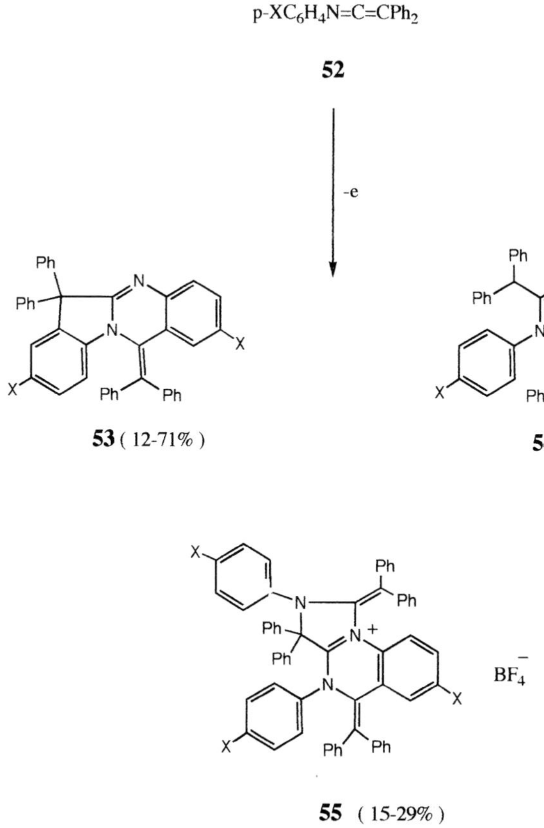

p-XC$_6$H$_4$N=C=CPh$_2$

52

−e

53 (12-71%)

54 (2-36%)

55 (15-29%)

BF$_4^-$

Scheme 28.

cycloaddition is preferred with the C=S bond because it is more polar than the C=N bond, so, accordingly, products of type **51** predominated.

The cyclic voltammetry of ketene imines of the type **52** (X = H, p-CH$_3$, p-CH$_3$O, p-Br) exhibited two irreversible waves in dichloromethane at a Pt-anode between 0.90–1.25 and 1.63–2.0 V vs Ag/AgCl, respectively. The controlled potential electrolysis at the first wave gave tetracyclic (**53**), bicyclic (**54**), and tricyclic (**55**) products [83] (Scheme 28).

Scheme 29 describes a plausible mechanism for the formation of the products which fit the observed coulometric (n ~ 0.45 F/mol) and preparative results. The intramolecular cyclization process involves a dimerization between a radical cation **52a**$^{\cdot+}$ and the ketene imine **52** to form the intermediate radical cation **52b** which then cyclizes to the radical **52c** which can abstract a hydrogen atom leading to **54** or can be further oxidized and transformed through a cyclization and deprotonation reaction to **53** which involves 1 F/mol. However, it seems that the [2 + 3]-cycloaddition between the parent compound **52** and the cation **52d** giving rise to **55** is the fastest reaction as compared with the intramolecular cyclization of **52d** to **53**. This can also explain the low consumption of electricity.

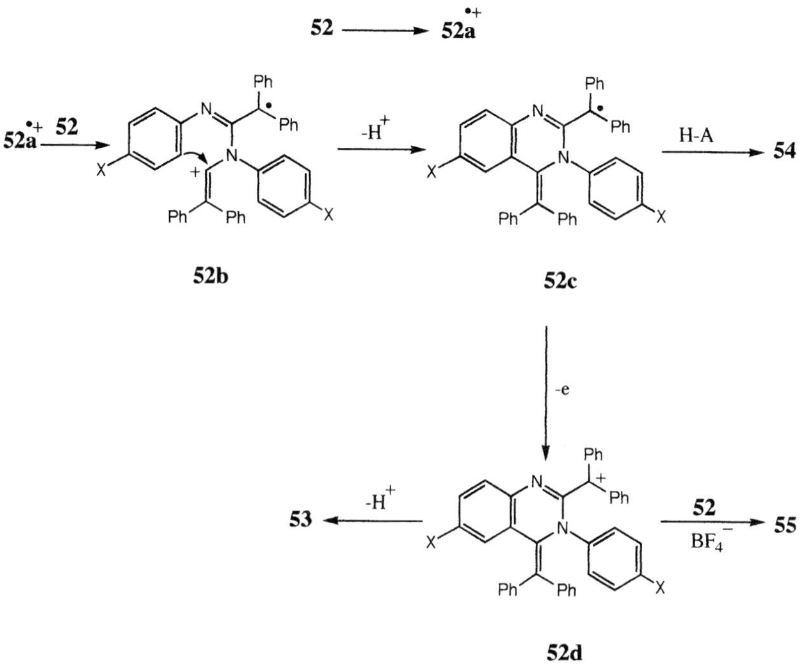

Scheme 29.

3.1.5 Oxidation of Amides

Steckhan and coworkers found that the indirect anodic oxidation of N-protected dipeptide esters **56**, in which the C-terminal amino acid is α-branched, can afford methyl imidazolidin-4-one-2-carboxylate **57** in 45–84% yields [86]. This reaction can be performed at a Pt-anode by using Et_4NCl as an electrolyte in the presence of 5% methanol in an undivided cell (Scheme 30).

The structure of the acid moiety of an amide can have a large influence on the reactivity of the amide. Shono and coworkers [87] found an efficient method for the preparation of pyrrolidines from amides. The indirect anodic oxidation of N-alkyltosylamides **58** in a two phase system containing KOH and KBr under heating gave pyrrolidines in 86 to 100% yield (Scheme 31). The reaction route for the formation of **59** seems to be analogous to the Hofmann-Löffler reaction. Thus, four consecutive reaction steps involving homolytic cleavage of the N–Br bond of **58** to yield radical **58b** and Br·, subsequent abstraction of δ-hydrogen of **58b**, coupling of the radicals **58c** and Br·, and base-induced ring closure reaction of the δ-halo intermediate **58d** yield the final product **59**.

Anodic oxidation of benzoylacetanilide in CH_3CN-0.1 mol/l $LiClO_4$ solution using the controlled potential electrolysis (E = 1.8 V vs Ag/AgCl) gave 2,4-dioxo-1-phenyl-1,2,3,4-tetrahydroquinoline **61** [88] (Scheme 32). The reported yield (75%) of the product **61** is surprisingly high since at least two different moieties, i.e., enol and amide, are oxidizable at the applied potential, which could result in a complex mixture of the products. The authors proposed a mechanism for the cyclization reaction involving the coupling between an N-centered radical and the radical cation of the benzene ring, which is probably an incorrect assumption. Namely, an N-centered radical would easily be further

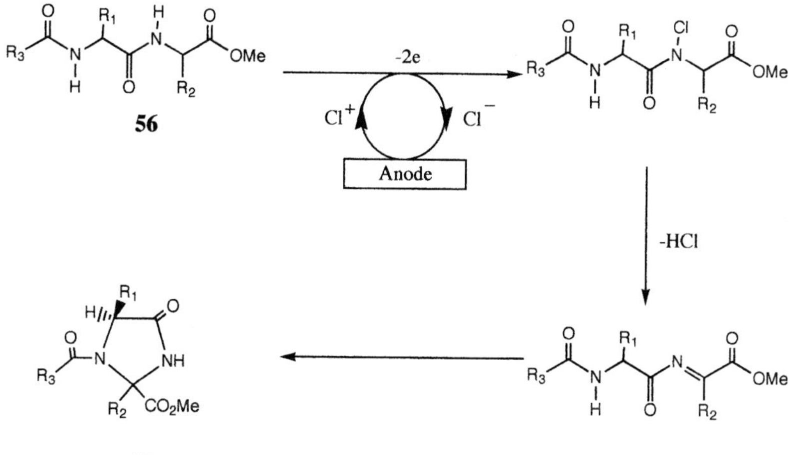

Scheme 30.

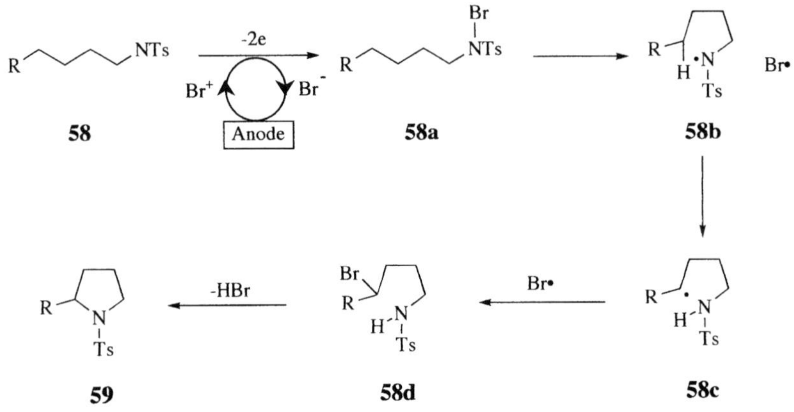

Scheme 31.

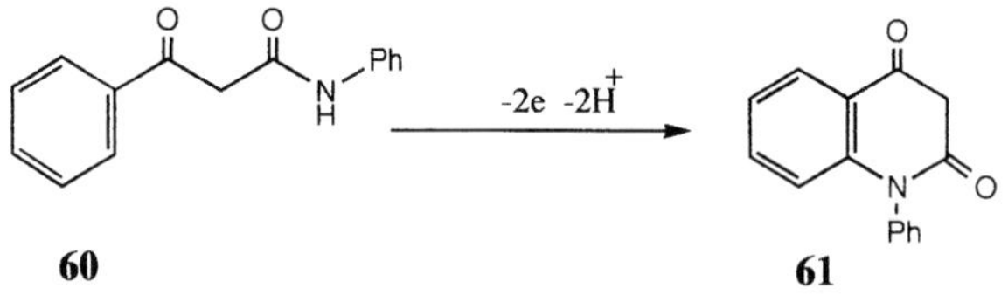

Scheme 32.

oxidized at the applied potential. Notably, the generation of the benzene radical cation substituted with an electron-accepting carbonyl group would require a higher oxidation potential.

3.1.6 Oxidation of Activated Aromatic Rings

The selective oxidation of the activated aromatic ring, substituted with electron-donating hydroxy or methoxy groups, can be perfomed at relatively low electrode potential (E_p = 0.3–1.2 V vs SCE) and ring closure is the result of the intramolecular nucleophilic attack of an amino group on the oxidized aromatic ring.

Swenton and coworkers developed a useful synthetic method for the preparation of quinone imine ketals [89, 90]. Anodic oxidation of N-trifluoroacetyl derivatives **62** in 2% KOH/CH_3OH solution (R = CH_3O, Br) or in 2% $LiClO_4/CH_3OH$ solution (R = H) in a divided cell gave bisketal **63**. Two consecutive reactions during the work-up procedure involving hydrolysis of **63** to **64** and a condensation reaction gave quinone imine ketals **65** in 40–91% yield (Scheme 33).

Zhang and Dryhurst studied the electrochemical and enzyme-mediated oxidation of tetrahydropapaveroline **66** with the objective of investigating the oxidation chemistry and isolating the major oxidation products [91]. The

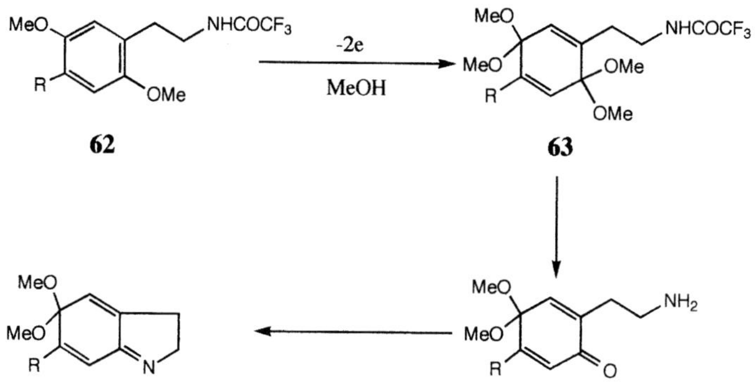

Scheme 33.

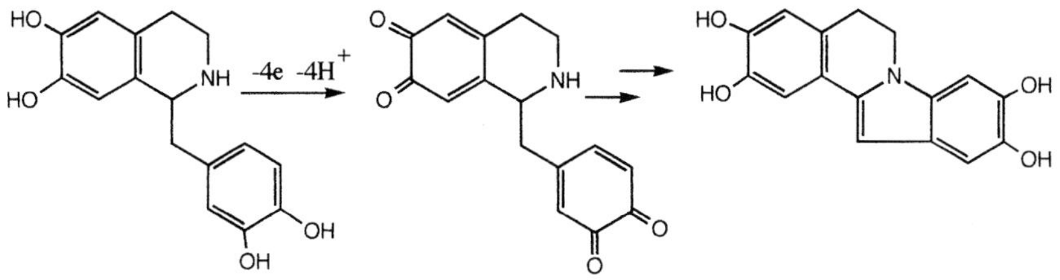

Scheme 34.

alkaloid **66** is the metabolic product of the neurotransmitter dopamine and it was suggested that **66** might have addictive liability or be further metabolized to morphine-like compounds which are responsible for the syndrome of alcohol addiction. Controlled potential oxidation of **66** at E = 0.44 V vs SCE in pH 3.0 phosphate buffer gave compound **68** as the major product during the first few minutes of the electrolysis. Since **68** is oxidized at a more negative potential (Ep = 0.38 V vs SCE at pH 3.0), it is subsequently oxidized in parallel with **66**. The initially formed quinone **67**, following the appropriate rearrangements, is transformed to **68** (Scheme 34).

3.2 Formation of the Carbon–Oxygen Bond

3.2.1 Oxidation of Endiamines and Enaminones

Anodic oxidation of endiamine **69** provided an example of double cyclization to an indolo-oxazoline **70** [92]. The oxidation was carried out in CH_3CN-

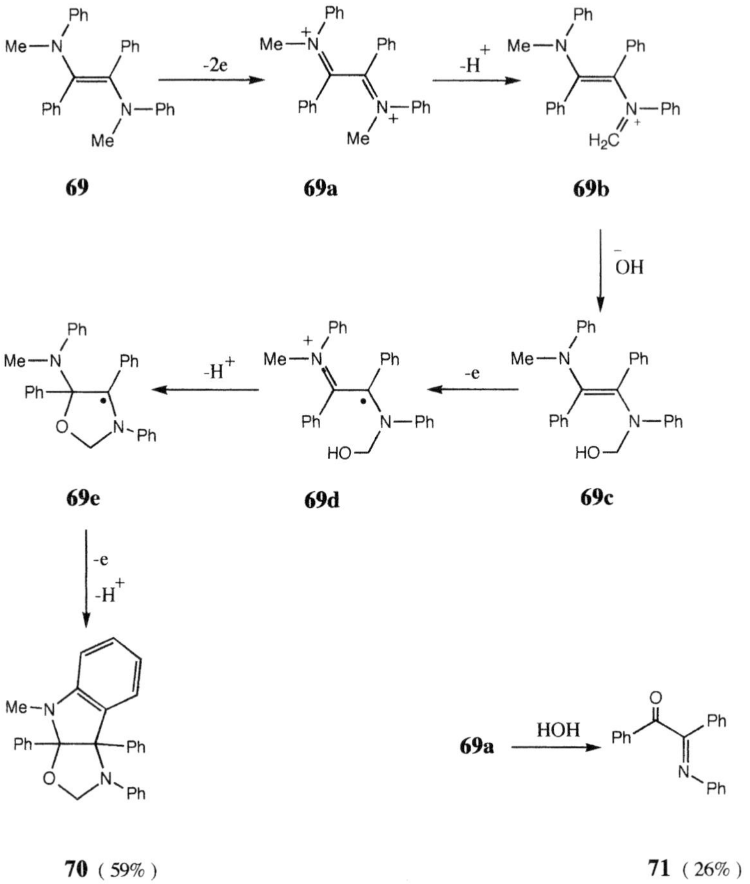

Scheme 35.

0.1 mol/l KPF$_6$ solution in the presence of 2,6-lutidine as a base using controlled potential electrolysis (E = 0.4 V vs SCE). Scheme 35 illustrates the pathway to the unexpected formation of **70**. Endiamine **69** is oxidized to dication **69a**, which is then deprotonated by base giving rise to cation **69b**. The attack of water leads to **69c** which is readily oxidized at the applied potential to radical cation **69d**. Double cyclization at the anode yields **70** as a major product. The formation of the by-product **71** arises probably by partial hydrolysis of dication **69a**.

The mechanism for the transformation of enaminones **1** via anodic oxidation to bischromones **72** was discussed in Sect. 2.1. Since the enaminones can be easily prepared from available o-hydroxyacetylarenes the reaction shown in Scheme 36 represents a general method for the synthesis of the corresponding bischromones which can be obtained in good yields [21].

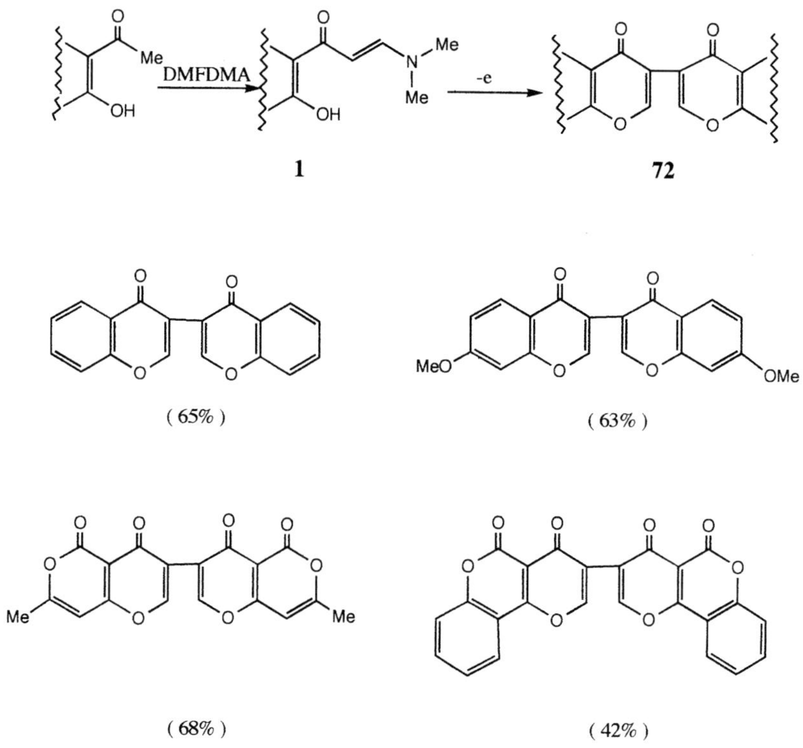

Scheme 36.

3.2.2 Oxidation of N-Acylhydrazones and 1-Arylmethylenesemicarbazides

The oxidative cyclization of aldehyde N-acylhydrazones 73 (R_2 = H) in methanolic sodium acetate solution afforded the corresponding 2,5-disubstituted 1,3,4-oxadiazols 74 in yields of between 30 and 77%. The oxidation of ketone N-acylhydrazones (R_2 = H) gave 2-methoxy-Δ^3-1,3,4-oxadiazolines 75 in yields of 50–60% [93]. The anodically induced cyclization appears to involve a cationic intermediate 73a through the loss of the two electrons and one proton (Scheme 37). The cationic center of azomethynyl carbon would be attacked intramolecularly by the carbonyl oxygen leading to 73b or 73c. In the case of aldehyde N-acylhydrazones the product of such an attack can lose a proton and rearrange to form the stable oxadiazole 74. However, in the case of ketone N-acylhydrazone, the nucleophilic attack of methanol gave rise to the oxadiazolines 75 as the final products of the reaction.

The oxadiazoles 77 were also obtained by anodic intramolecular cyclization of 1-arylmethylenesemicarbazides 76 in CH_3OH-0.1 mol/l Et_4ClO_4 solution in 51–65% yield [55]. The aromatic methyl esters were also formed either as a minor or major product in 55–80% yield (Scheme 38). The selectivity of the

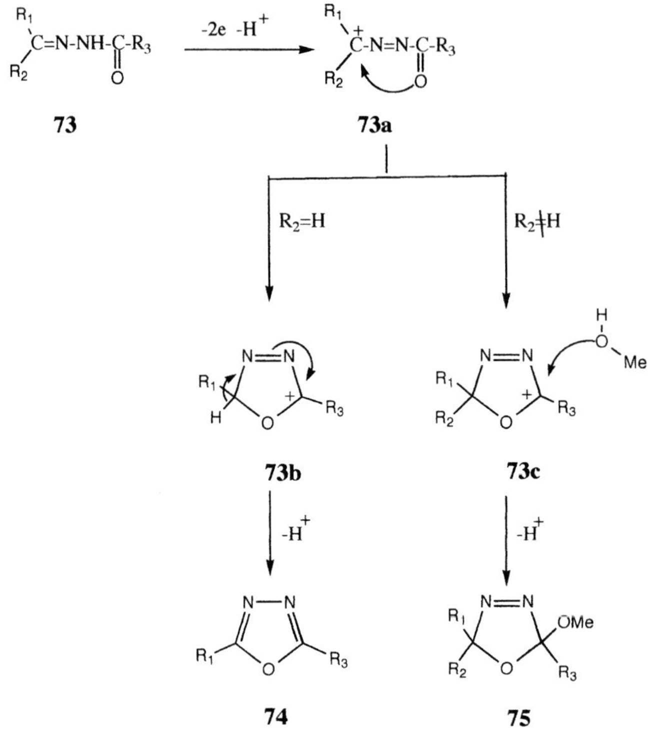

Scheme 37.

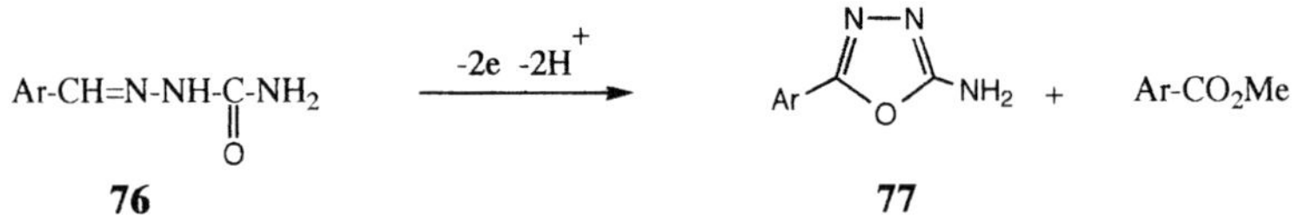

Scheme 38.

anodic oxidation of **76** in methanol, i.e., the formation of oxadiazole vs aromatic methyl esters, seems to be dependant on the donating capability of the aryl group. The anodic oxidation of **76** with stronger electron-donating groups (Ep of **76** < 1.2 V vs SCE) gave rise to the formation of oxadiazols **77** as major products while weaker electron-donating groups (Ep of **76** > 1.2 V vs SCE) led to the formation of esters as major products.

3.2.3 Oxidation of Enols and Olefins

Surprisingly little preparative work has been done on the anodic oxidation of enols and enolates, although the resulting α-carbonyl radicals are important intermediates in synthetic organic chemistry and biological systems [94]. Due to the high concentration of the anodically generated α-carbonyl radicals in the diffusion layer, the major products in the anodic oxidation of enolates are dimers [95–97]. However, Schäfer and Al Azrak demonstrated that the generated α-carbonyl radicals add to the double bond of an electron-rich olefin present in excess and the resulting radical can either dimerize or further be oxidized to the α-carbonyl cation [98]. Torii et al. designed experimental conditions for the anodic synthesis of dihydrofuran derivatives [99] through the oxidation of 1,3-cyclohexanedione in the presence of ethyl vinyl ether. A similar reaction concept was used for the synthesis of hexahydrobenzofuran derivatives by Yoshida and coworkers [100]. They investigated the oxidation of 1,3-diketones **78** (1,3-cyclohexanedione, dimedone, 1,3-cyclopentanedione, 2-methyl-1,3-cyclopentanedione) in the presence of olefins (styrene, isoprene, butadiene, ethyl vinyl ether, allyltrimethylsilane) in CH_3CN-0.2 mol/l Et_4NOTs solution and the corresponding heterocycles **79** were obtained in 45–90% yield. The mechanism rationalizing the formation of the products **79** is shown in Scheme 39. The oxidizable species is the enol form of **78** since its potential is usually 1 V more negative than that of the parent ketone (see [94], p. 211). The radical cation **78a** deprotonates to the α-carbonyl radical **78b** and adds to the olefin to produce radical **78c**. This is oxidized at the anode to form the cation **78d** which cyclizes through intramolecular attack of the carbonyl group to give **78e**. β-Proton loss yields the final dihydrofuran derivative **79**.

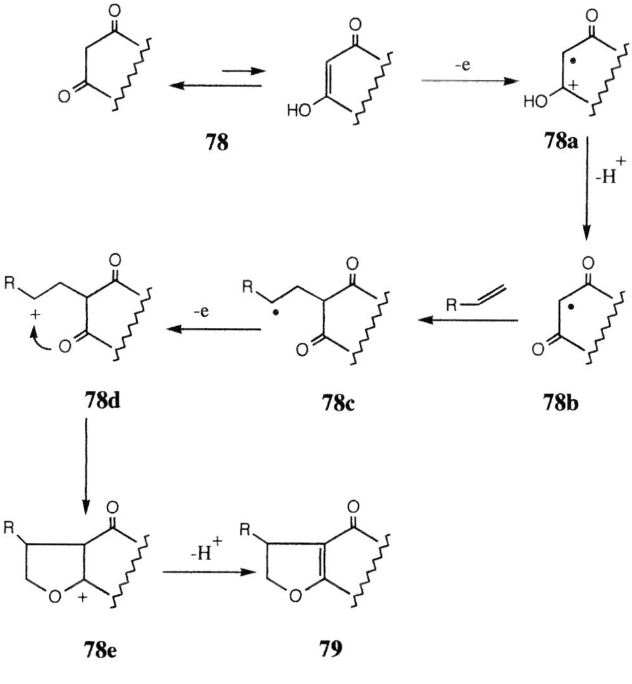

Scheme 39.

The anodic oxidation of 1,3-diketones in the presence of olefins in an oxygen atmosphere gave the extremely stable cyclic peroxides **80** in good yield [101, 102] (Scheme 40). A catalytic amount of electricity was sufficient for the reaction and an electro-intiated radical chain mechanism was suggested.

It has been found that the electrochemically generated NO_3^- radical addes to the substituted olefins **81**, and the radical species **81a** formed is further oxidized to the cationic intermediate **81b** which reacts with acetonitrile and yields **82** (Scheme 41). The anodic oxidation was carried out in a mixed solvent CH_3CN-Et_2O with $NaNO_3$ as a supporting electrolyte. The oxazoline derivatives **82** were isolated in 69–77% yield [103].

We have studied the dependence of the oxidation peak potentials on the substitution in the rings A and B for 16 structurally different 2-hydroxychalcones [61]. The increase in the number of electron donating groups in ring B resulted in the decrease of the peak potentials while the peak potentials remain virtually the same when the number of electron-donating groups in ring A was

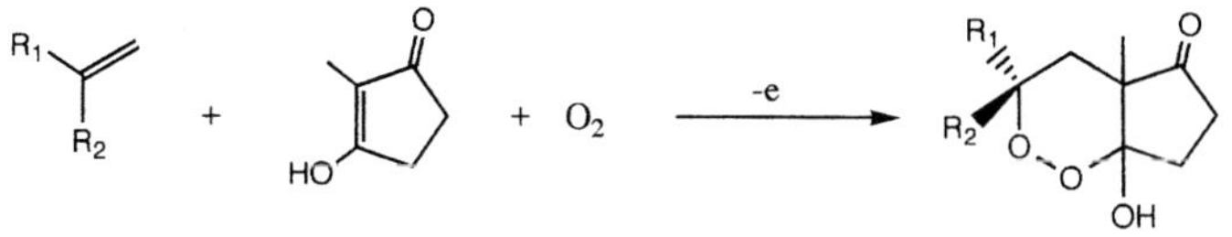

Scheme 40.

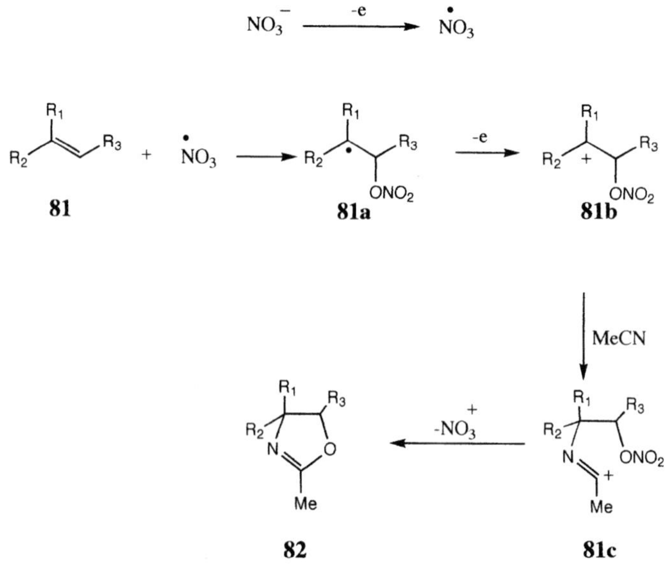

Scheme 41.

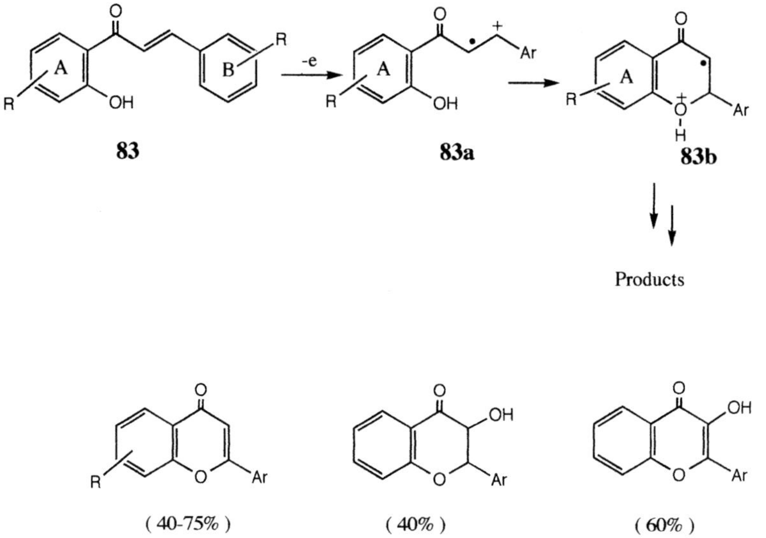

Scheme 42.

increased. One might propose that the oxidation of **83** is initiated by removal of the electron from the styrene moiety, thereby leading to radical cation **83a**, which then cyclizes to **83b** and, through different reaction pathways, is transformed to the final products (Scheme 42). The controlled potential electrolysis gave structurally different flavanoids in good yields depending on the structure of the substrate. Three types of products were formed: flavanones and 2,3-dihydroflavanols as two-electron oxidation products, and flavanols as four-electron oxidation products.

3.2.4 Oxidation of Alcohols and Catechol

Intramolecular cyclization of unsaturated alcohols to cyclic phenylselenoethers by means of organoselenium reagents has become an important tool for the synthesis is some natural products [104]. Instead of these chemical methods, a halide ion-promoted electrochemical oxyselenation of olefins to oxyselenides has been developed [105]. The intramolecular oxyselenation of unsaturated alcohols was performed by indirect oxidation of diphenyl selenide by anodically generated Br^+ or Br_2 [105–108]. The reactive intermediate ($PhSe^+$) reacts with the double bond of the unsaturated alcohol or the carboxylic acid giving rise to cation **84a** which cyclizes to the final products (Scheme 43). The products, found in good yields (50–80%), were five- or six-membered cyclic ethers or lactones depending upon the nucleophile present. The results obtained also showed that the substituents at the double bond and at the carbinol carbon atom have pronounced influence on the regioselectivity of the ring closure which was mainly in agreement with the Markovnikov rule. The stereoselectivity in some

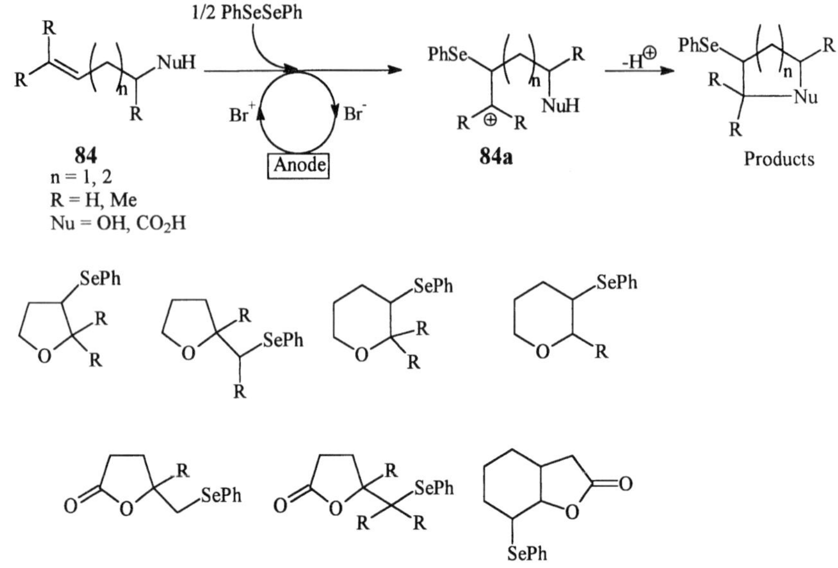

Scheme 43.

transformations appears to be very good. Similarly, intramolecular sulphoetherifications and sulpholactonizations of alkenols an alkenoic acids can be initiated by arylsulphenyl cations generated by direct or bromide mediated indirect anodic oxidation of diaryldisulphides [109].

We have studied the anodic oxidation of unsaturated alcohols using the controlled potential electrolysis (E = 1.9 V vs SCE) in CH_3CN-0.1 mol/l Et_4NClO_4 solution in a divided cell [110]. The oxidation of 4-pentenol after consumption of 0.8 F/mol gave 2-methyltetrahydrofuran and tetrahydropyran as the major products. The oxidation of 5-pentenol gave 2-methyltetrahydropyran and oxepam, while the oxidation of 3-butenol under the same reaction conditions did not give the cyclic products. We rationalized this reaction as the electrongenerated acid (EGA) catalyzed intramolecular cyclization (Scheme 44).

Anodic oxidation of 2-phenylethanol or 3-phenylpropanol at a Pt-anode in CH_3CN-0.1 mol/l Et_4NClO_4 solution using controlled potential electrolysis (E = 2.1–2.2 V vs SCE) gave low yields of the corresponding cyclic ethers [111], i.e., chroman (6%) and coumaran (15%), since in this case the competing β-fragmentation process is the major reaction pathway (Scheme 45). There are two possible sites within 2-phenylethanol **85** from which the electron may be removed, namely the electrons of the aromatic ring and the lone pair electrons on the oxygen atom. Using the equation given by Miller et al. [112] we were able to calculate the ionization potential for **85** (8.78 eV) which is quite close to the IP-value for toluene (8.82 eV). Since ethanol has an IP-value of 10.5 eV, it was reasonable to assume that the first electron was removed from the aromatic ring leading to the radical cation **85a** which cyclizes to radical **85b** and through

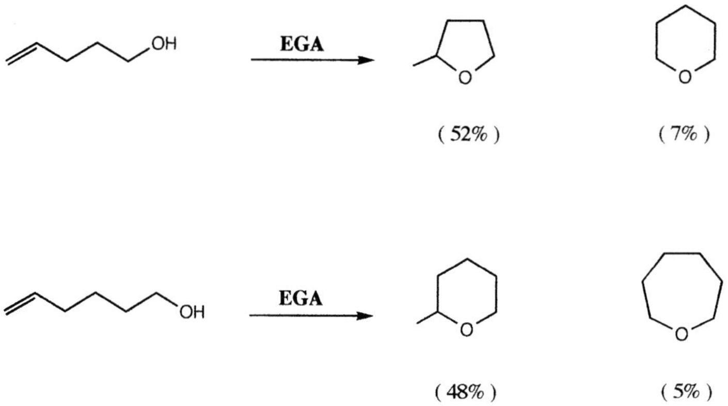

Scheme 44.

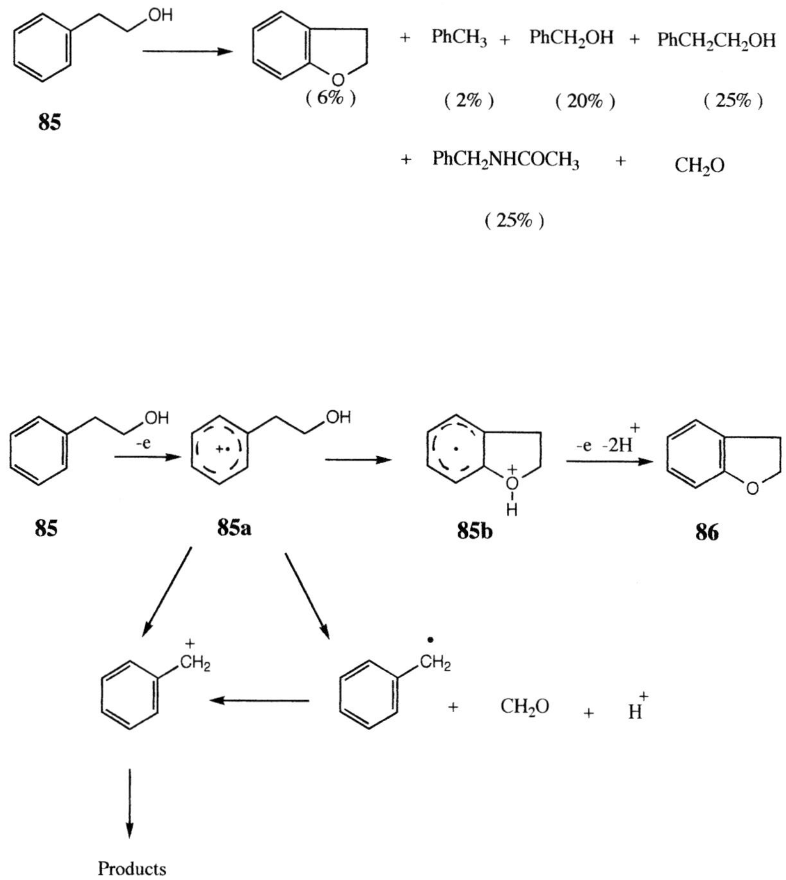

Scheme 45.

further deprotonation and oxidation reaction gives chroman **86**. It seems that the fragmentation of the radical cation **85a** is faster than the cyclization step. Since the benzyl radical formed has an IP-value of 7.76 eV, it is reasonable to propose that the benzyl radical is being oxidized at the applied potential to the benzyl cation which reacts with the nucleophile present (water or CH_3CN) leading to the final products.

Mechanistic aspects of the intermolecular cyclization reaction in the anodic oxidation of catechol in the presence of 4-hydroxycoumarin were discussed in Sect. 2.2. This reaction is a synthetically simple and versatile method for the preparation of formally [3 + 2] cycloadducts between a β-diketo compound and catechol [44, 45]. Anodic oxidation of catechol using controlled potential electrolysis (E = 0.9–1.1 V vs SCE) or constant current electrolysis (i = 5 mA/cm^2) was performed in water solution containing sodium acetate (0.15 mol/l) in the presence of various nucleophiles such as 4-hydroxycoumarin,

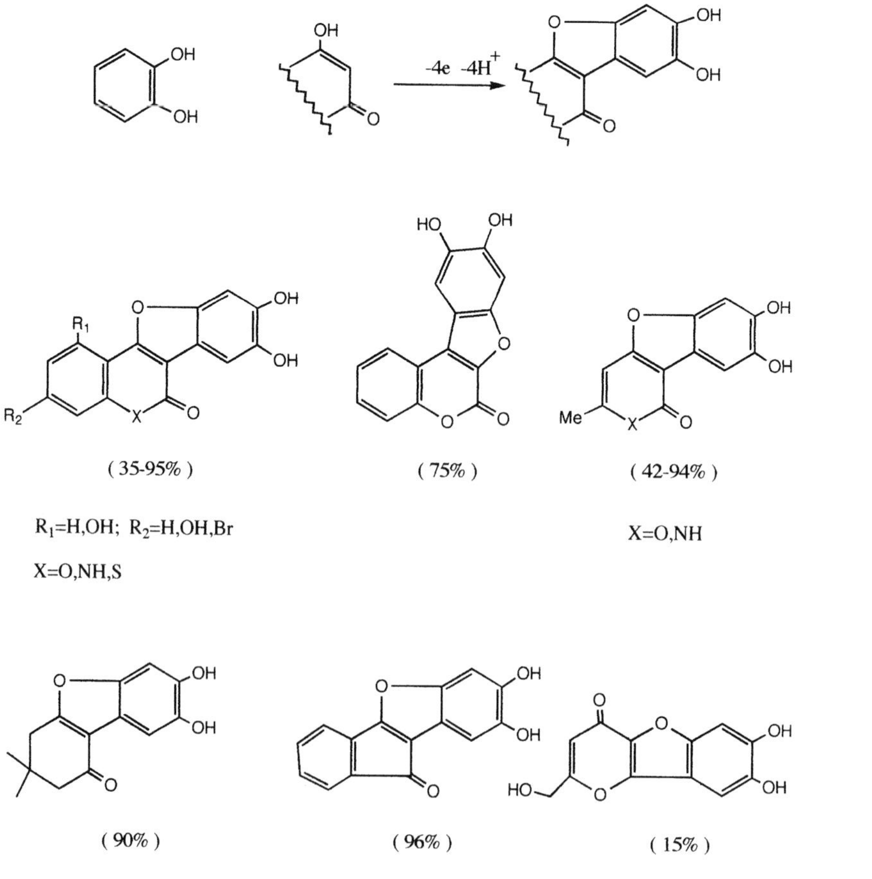

Scheme 46.

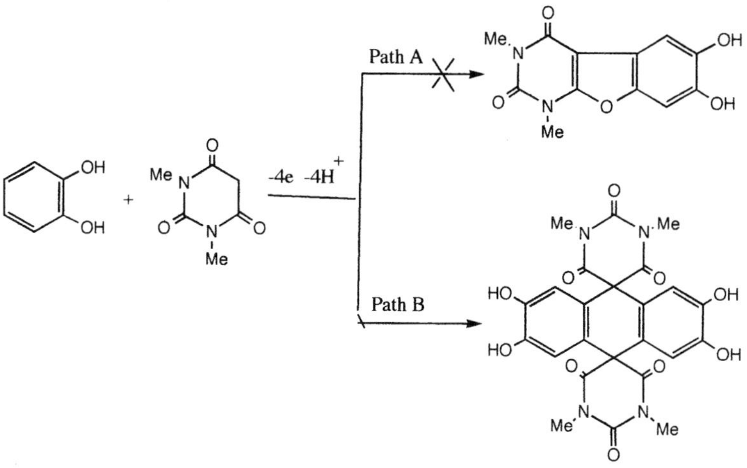

Scheme 47.

dimedone, 1,3-indandione, 4,7-dihydroxycoumarin, 4,5,7-trihydroxycoumarin, 4-hydroxy-6-bromocoumarin, 3-hydroxycoumarin, 4-hydroxy-6-methyl-α-pyrone, 4-hydroxy-6-methyl-2-pyridone, and 4-hydroxycarbostyrile. All electrochemical syntheses were carried out in an undivided cell at a graphite anode and a Pt-cathode leading to high yields of coumestan derivatives and related heterocyclic systems (Scheme 46). The oxidation potential of the heterocyclic system formed is more negative than the applied electrode potential but overoxidation was circumvented by precipitation of the products during the anodic oxidation. Recently Raju and coworkers reported the synthesis of an important drug by using the oxidation of catechol in the presence of kojic acid as the key step [113]. The product was obtained in 15% yield, while the oxidation of catechol in the presence of 4-hydroxythiocoumarin gave the corresponding cycloadduct in 35% yield [113].

The anodic oxidation of catechol in the presence of 1,3-dimethylbarbituric acid was carried out in aqueous solution containing sodium acetate in an undivided cell at graphite and nickel hydroxide electrodes [114]. The results did not fit with the expected structure (Scheme 47, path A) but a dispiropyrimidine was isolated in 35% yield (Scheme 47, path B). It seems that the initial attack of 1,3-dimethylbarbituric acid on the anodically formed o-quinone does not occur through the carbon–oxygen bond formation but rather through the carbon–carbon bond formation, giving rise to the final product via several consecutive reaction steps.

3.2.5 Oxidation of Amides and Sulfides

The anodic oxidation of amides and carbamates is one of the most general and efficient electrochemical reactions known [115]. The primary intermediate

N-acyliminium ion can be trapped by methanol to afford α-methoxy amides or by an intramolecular nucleophilic center giving rise to heterocyclic products [116, 117]. Moeller et al. oxidized the dipeptide **87** (Boc-Hse-L-Pro-OMe) in a 95:5 mixture of acetonitrile/isopropyl alcohol using 0.1 M Bu$_4$NBF$_4$ as supporting electrolyte in an undivided cell, with two platinum electrodes and constant current [118]. The bicyclic dipeptide **88** was formed, through the *N*-acyliminium ion **87a** generated in situ, in 48% yield (Scheme 48). The reaction was regioselective and only one (i.e., **87a**) of three possible *N*-acyliminium ions was generated. The isomer **88** having an *S*-configuration at the ring fusion dominated the product ratio by more than 15:1. It seems that the cyclization is not influenced by the chirality of the α-carbon of the proline precoursors and, therefore, must be dictated by the chirality of the homoserine residue.

In the course of developing a new synthetic methods for the construction of chromone and spirochromone skeletons, Chiba et al. used anodic oxidation to generate the reactive *o*-quinone methides which are trapped by unactivated alkenes to form chromones including euglobal skeletons [119]. The controlled potential anodic oxidation in CH$_3$NO$_2$-0.5 mol/l LiClO$_4$ solution of the compound **89** gave radical cation **89a** which is transformed through loss of the

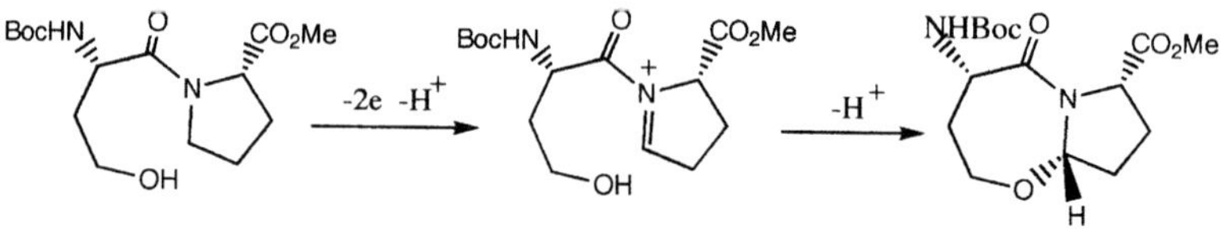

Scheme 48.

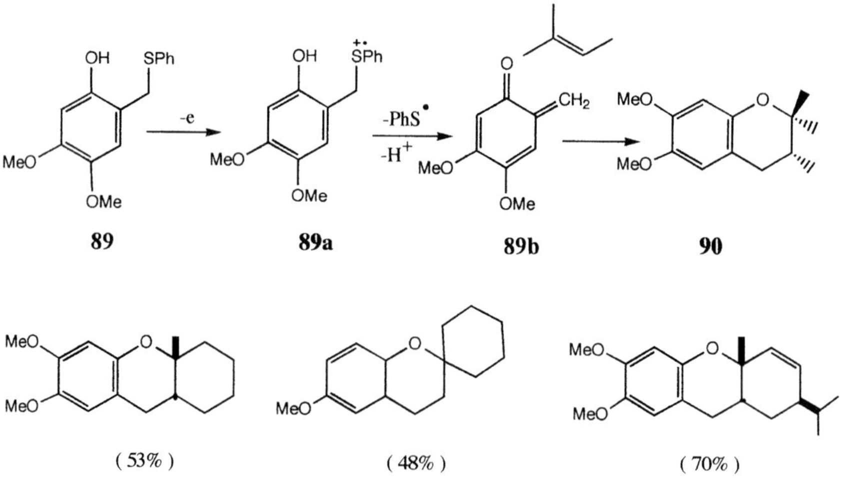

Scheme 49.

phenylthio radical and a proton to *o*-quinone methide **89b** which reacts intermolecularly with the olefin present to give cycloadduct **90** in 77% yield (Scheme 49). These cycloaddition reactions were stereo- and regioselective to form the *cis* adducts in which the ether oxygen was attached to the C-1 position. Some of the products shown in Scheme 49 were formed in 48–74% yield.

3.3 Formation of the Carbon–Sulfur Bond

Only a small number of examples of the cyclization reactions through the carbon–sulfur bond formation involving the anodic oxidation of thiobenzanilide, thioactetanilide [120], thiocarboxanilide [121], and the conversion of disulfides to various 2-*β*-methyl-substitued penicillins [122] have been reported. 2-Aminobenzothiazoles were prepared by the reaction of electrogenerated thiocyanogen with aromatic amines in 36–92% yield [123]. Anodic oxidation was carried out in aqueous acetic acid containing 0.8 mol/l ammonium thiocyanate at 5–10 °C.

3.4 Formation of the Nitrogen–Nitrogen Bond

3.4.1 Oxidation of Hydrazones and Formazans

A novel way to preparing *s*-triazolo[3,4-*a*]pyridinium salts **92** by anodic oxidation of arylhydrazones **91** of 2-acetylpyridine, 2-benzoylpyridine, and formylpyridine in CH_3CN-Et_4NX (X = ClO_4, *p*-TsO, BF_4) has been reported [51]. Electrochemical oxidative cyclization (Scheme 50) is superior to chemical oxidation [124]: (i) the products are obtained in high yield (79–91%); (ii) pyridinium salts with different anions can be prepared by using a supporting electrolyte carrying the desired anion; (iii) the configuration of the substrate does not affect the yield of the product, e.g., both *syn* and *anti* **91** gave **92** in high yield, while in the chemical oxidation only *syn*-hydrazone gave the cyclic product; (iv) products are formed by oxidation of arylhydrazone of 2-formylpyridine (R = H) which was not the case in the chemical oxidation.

Anodic oxidation of the monooxime phenylhydrazone of a 1,2-dicarbonyl compound **93** in CH_3CN-0.1 mol/l Et_4NClO_4 solution gave 2-phenyl-1,2,3-triazol-1-oxide **94** in very good yield [125, 126] (Scheme 51).

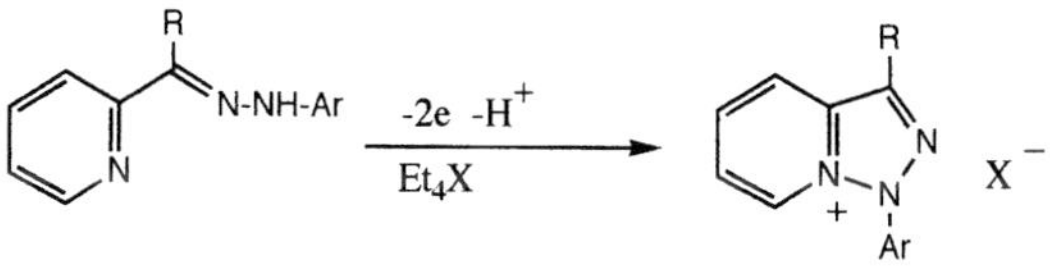

Scheme 50.

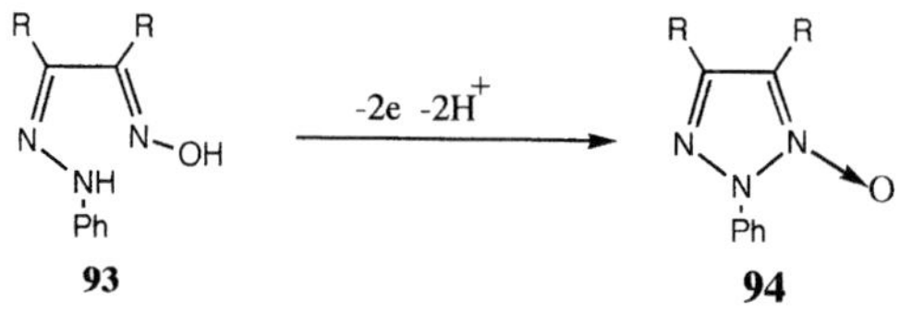

Ibro Tabaković

$$\underset{93}{\underset{\underset{Ph}{|}}{\underset{NH}{N}}\overset{R}{\underset{}{\diagdown}}\overset{R}{\underset{}{\diagup}}N\text{-OH}} \xrightarrow{-2e\ -2H^+} \underset{94}{\underset{\underset{Ph}{|}}{N}\overset{R}{\underset{N}{\diagdown}}\overset{R}{\underset{N}{\diagup}}O}$$

Scheme 51.

The chemical oxidation of 1,3,5-triarylformazans to tetrazolium salts was first accomplished in 1894 [127]. Almost no attention was given to these compounds for about 50 years after their discovery. This situation began to alter markedly because of the application of tetrazolium salts in histochemical, pharmacological, and other biomedical research areas [128]. Specifically, the tetrazolium salt is reduced to a colored formazan derivative by reducing enzymes found only in metabolically active cells. Anodic transformation of formazans to tetrazolium salts was performed in acetonitrile solution using cotrolled potential electrolysis [17, 129]. In our view this reaction could be considered as a method of choice for the preparation of tetrazolium salts. The products were obtained in high yield and the electrolysis can be performed in a divided cell under constant current and decoloration of the solution indicates the end point of the reaction. Recently the anodic oxidation of formazans to tetrazolium salts was performed successfully in aqueous ethanol solution [130].

3.4.2 Oxidation of Triazenes

Speiser and coworkers have studied the anodic oxidation of various triazenes in view of their importance as chemotherapeutics against malignant melanoma [131–134]. Trizenes are oxidized to the radical cation **95a** whose stability strongly depends on the *para* substitution in the aromatic ring. The decay of **95a** is fast for weaker electron-donating substituents through cleavage of the N(2)–N(3) bond in respective cations **95b** and radicals **95c** (Scheme 52).

The existence of diazonium ion **95b** is confirmed by controlled potential oxidation of **95** in CH_3CN-0.1 mol/l Bu_4NPF_6 solution in an undivided cell [134]. The diazonium ion **95b** reacts with the cathodically formed 3-aminocrotononitrile in a paired-type electrolysis to form enamine **96** which undergoes either hydrolysis to give **98** or oxidative intramolecular cyclization to give triazole **97** (Scheme 53).

3.5 Formation of the Nitrogen–Oxygen Bond

3.5.1 Oxidation of Vicinal Dioximes

In 1978 we reported the first anodic oxidation of benzyldioxime which afforded 3,4-diphenylfuroxan in good yield [125]. Chemical oxidation of vicinal dioximes

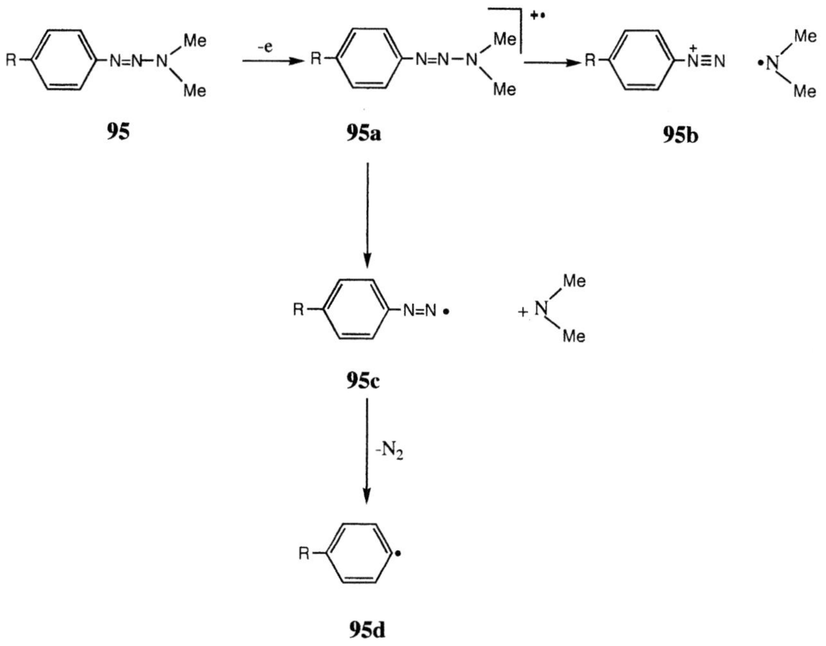

Scheme 52.

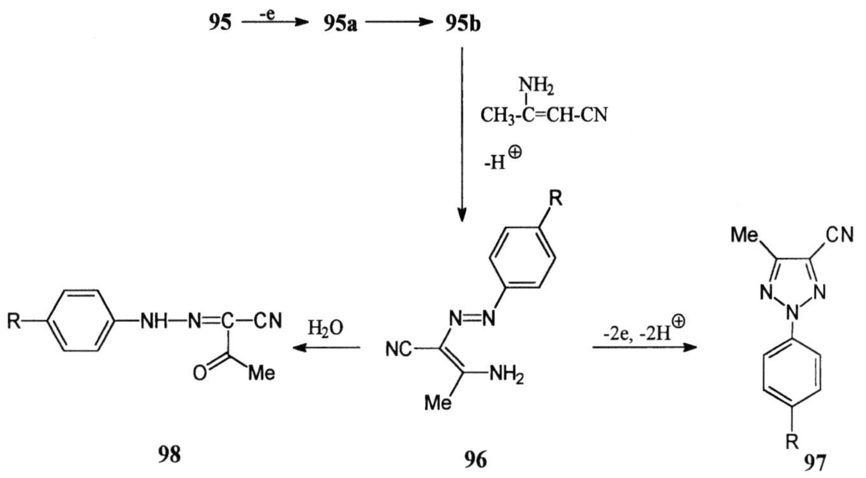

Scheme 53.

is one of the most common methods for the synthesis of furoxanes, which are energy rich materials of practical importance. However, is disadvantageous because of chemical oxidation fire and explosion hazards of the process, toxicity and high cost of the reactants, and often insufficient selectivity and yields of the

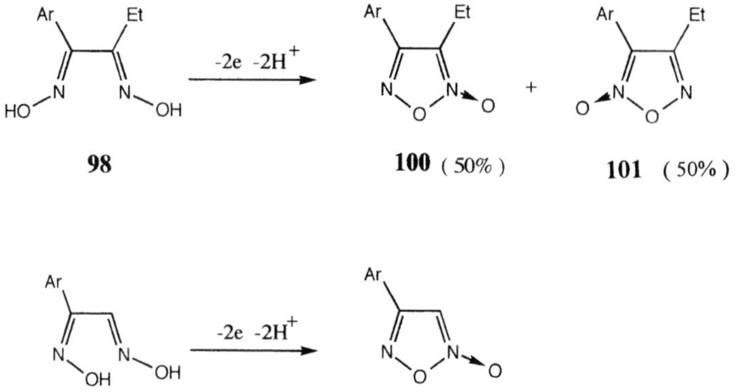

Scheme 54.

process [135]. The electrochemical alternative appeared to be a suitable method for the synthesis of furoxanes under mild reaction conditions. Jugelt et al. [126, 136] and Niyazymbetov et al. [137–141] have thoroughly studied the most important aspects of this reaction.

Jugelt et al. [136] studied the regioselectivity of the anodic oxidation of *anti*-**98** and *amphi*-**99** vicinal dioximes on the structure of furoxanes obtained after anodic oxidation under constant current. The furoxanes (**101, 102**) were obtained in 22–65% yield but the ratio of the isomeric furoxane was dependent on the structure of the substrate (Scheme 54).

The mechanism of the reaction following an **e-p-e-C-p** scheme was correctly envisaged by Jugelt et al. [136] and proved experimentally by Niyazymbetov et al. [140]. The selected vicinal dioximes, studied by RDE and RRDE voltammetry in CH_3CN-0.1 mol/l $LiClO_4$ solution, showed two diffusion controlled waves. The first wave ($E1/2 = 1.27$–1.82 V vs Ag/Ag^+) corresponded to the oxidation of the vicinal dioxime **103** and the second wave ($E1/2 = 1.69$–2.55 V vs Ag/Ag^+) to the oxidation of the furoxane formed along the first wave. Using RRDE voltammetry the primary radical cation **103a** was generated on the disc and the products were analyzed on the ring. The current-potential curve obtained on the ring showed one anodic wave at $E1/2 = 0.42$–0.45 V vs Ag/Ag^+ which corresponded presumably to the oxidation of iminoxyl radical **103b** to the oxiimmonium cation **103c** and a cathodic wave with $Ep/2 = -0.5$ V vs Ag/Ag^+ which corresponded to the reduction of the protons liberated during the anodic oxidation. The controlled potential coulometry at the plateau of the first anodic wave showed that the ovrall electrode reaction is a two-electron oxidation. The proposed mechanism is shown in Scheme 55.

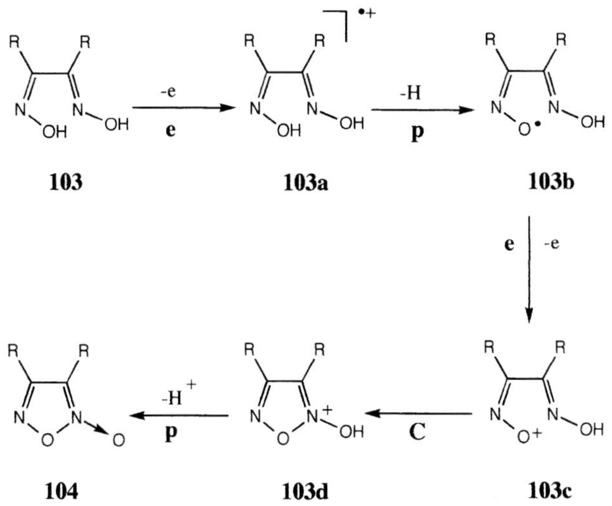

Scheme 55.

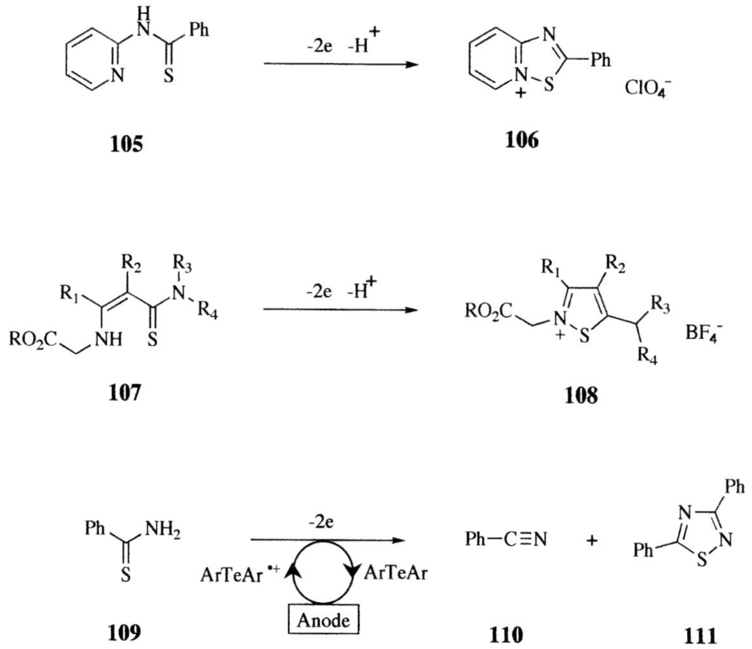

Scheme 56.

In order to find the best reaction conditions for the electrosynthesis of furoxanes, Nyazymbetov et al. [141] have extensively studied all parameters which affect the yield of furoxane such as solvent, supporting electrolyte, electrode material, controlled potential vs constant current electrolysis, temperature, degree of conversion, and added base as a proton acceptor. The optimum yields of the reaction were between 90 and 96%.

3.6 Formation of the Nitrogen–Sulfur Bond

3.6.1 Oxidation of Thioamides

To the best of our knowledge there are only three papers dealing with the cyclization reactions occurring through nitrogen–sulfur bond formation. Oxidation of N-(2-pyridyl)thiobenzamide **105** at controlled potential in CH_3CN-0.1 mol/l Et_4NClO_4 solution gave 2-phenyl-1,2,4-thiadiazolo[2,3-a]pyridinium perchlorate **106** in 80% yield [120] (Scheme 56). Similar intramolecular anodic oxidation of thioacrylamides **107** afforded 5-amino-1,2-thiazolium salts **108** in 62–90% yield [142]. The 1,2-thiazolium salts obtained were easily transformed to 3-aminopyrrole derivatives. The indirect anodic oxidation of thioamide **109** with the generated radical cation of bis(p-methoxyphenyl)telluride (ArTeAr) as a mediator gave benzonitrile **110** and/or 1,2,4-thiadiazol **111** as the reaction products under various conditions [143]. Depending on the reaction conditions used 1,2,4-thiadiazoles can be selectively formed in yields between 90 and 98%.

Acknowledgement. I wish to express my sincere thanks to my collaborators from the University of Banjaluka in Bosnia, who are now living in USA, Canada, England, Sweden, Denmark and Germany. Their names are cited in the references. I am also grateful to Professor Larry L. Miller, University of Minnesota, for his encouragement, support and inspiring discussions.

4 References

1. (a) Lund H, Baizer MM (1991) Organic electrochemistry, 3rd edn. Marcel Dekker, New York; (b) Torii S (1985) Electroorganic synthesis: oxidations, methods and applications-part 1. Kodansha/Verlag Chemie, Tokyo/Weinhem; (c) Shono T (1984) Electroorganic chemistry as a new tool in organic synthesis. Springer, Berlin Heidelberg New York; (d) Shono T(1991) Electroorganic synthesis. Academic Press, New York
2. Miller LL, Kariv E, Behling JR (1977) Ann Rep in Med Chem 12: 309
3. Heinze J (1984) Angew Chem Int Ed 23: 831
4. (a) Lund H (1970) Adv Heterocyclic Chem 12: 213; (b) Baumgartel H, Retzlav KJ (1984). In: Bard AJ, Lund H (eds) Encyclopedia of electrochemistry of the elements, vol 15. Marcel Dekker, New York; (c) Lund H, Tabakovic I (1984) Adv Heterocyclic Chem 36: 235; (d) Armand J, Pinson J (1983). In: Gupta RR (ed) Physical methods in heterocyclic chemistry, vol 7. Wiley, New York; (e) Dryhurst G (1977) Electrochemistry of biological molecules. Academic Press, New York; (f) Toomey JE (1985) Adv Heterocyclic Chem 37: 167; (g) Lacan M, Tabakovic I (1974) Kem Ind (Zagreb) 23: 225; (h) Lacan M, Tabakovic I (1975) Kem Ind (Zagreb) 24: 227; (i) Tabakovic I (1977) Bull Soc Chim (Belgrade) 42: 761; (j) Simonet J, Le Guillanton G (1986) Bull Soc Chim Fr 221
5. Andrieux CP, Saveant JM (1974) J Electroanal Chem 53: 165; (b) Ammar F, Andrieoux CP, Saveant JM (1973) J Electroanal Chem 53: 407

6. Andrieux CP, Saveant JM, Tessier D (1975) J Electroanal Chem 63: 429
7. Anne A, Hapiot P, Moiroux J, Neta P, Saveant JM (1992) J Am Chem Soc 114: 4694
8. Amatore CM, Saveant JM, (1978) J Electroanal Chem 86: 227
9. Fry AJ, Little RD, Leonetti J (1994) J Org Chem 59: 5017
10. (a) Parker VD (1986) Topics in Organic Electrochemistry (Fry AJ, Britton WE, eds.) Plenum Press, New York, Chapter 2; (b) Aalstad B, Parker VD (1982) Acta Chem Scand B36: 187
11. Aalstad B, Ronlan A, Parker VD (1982) Acta Chem Scand B36: 199
12. Aalstad B, Ronlan A, Parker VD (1982) Acta Chem Scand B36: 171
13. Parker VD (1981) Acta Chem Scand B35: 259
14. Ahlberg E, Parker VD (1981) Acta Chem Scand 121: 73
15. Aalstad B, Parker VD (1980) Acta Chem Scand 112: 163
16. Andrieux CP, Delgado G, Saveant JM, Su KB (1993) J Electroanal Chem 348: 107
17. Tabakovic I, Trkovnik M, Grujic Z (1979) J Chem Soc Perkin 2: 166
18. (a) Umemoto K (1985) Bull Chem Soc Jpn 58: 2051; (b) Marques EP, Zhang J, Metcalfe RA, Pietro WJ, Lever ABP (1995) J Electroanal Chem 395: 133
19. Umemoto K (1989) Bull Chem Soc Jpn 62: 3783
20. Maender OW, Russell GA (1966) J Org Chem 31: 442
21. Sanicanin Z, Tabakovic I (1988) Electrochim Acta 33: 1601
22. Tabakovic I, Crljenak S (1981) Heterocycles 16: 699
23. Jugelt W, Grubert L (1985) Z Chem 25: 408
24. Jugelt W, Grubert L (1985) Z Chem 25: 442
25. Jugelt W, Grubert L (1985) Z Chem 25: 443
26. Grubert L, Jugelt W (1988) Z Chem 28: 187
27. Grubert L, Jugelt W, Bress HJ, Strietzel U (1990) Z Chem 30: 286
28. Grubert L, Jugelt W, Bress HJ, Strietzel U, Dombrowski A (1992) Liebigs Ann Chem 885
29. Gunic E, Tabakovic I (1985) J Org Chem 53: 5081
30. Warkentin J (1970) Synthesis 279
31. Butler RN (1984) Chem Rev 84: 249
32. Lin EC, VanDeMark R (1982) J Chem Soc Chem Commun 1176
33. Okimoto M, Chiba T (1990) J Org Chem 55: 1070
34. Wang Q, Amer A, Mohr S, Ertel E, Jochims JC (1993) Tetrahedron 44: 9973
35. Wang Q, Amer A, Troll C, Fischer H, Jochims JC (1993) Chem Ber 126: 2519
36. Shawali AS (1993) Chem Rev 93: 2731
37. Fukui K (1971) Acc Chem Res 4: 57
38. Barbay G, Caullet C (1974) Tetrahedron Lett 1717
39. Hammerich O, Parker VD (1972) J Chem Soc Perkin 1: 1718
40. Okimoto M, Chiba T (1990) J Org Chem 55: 1070
41. Chiba T, Okimoto M (1992) J Org Chem 57: 1375
42. Stankovic S, Agray G, Tabakovic I, Gunic E (1991) Acta Cryst 47: 1210
43. (a) DeLaHoz A, Pardo C, Sanchez A, Elquero J (1990) J Chem Res 294; (b) Ivanova VKh, Brezylin BI, Kitaev Yu (1977) Izv Akad Nauk SSSR Ser Khim 2: 393
44. Grujic Z, Tabakovic I, Trkovnik M (1981) Tetrahedron Lett 22: 4823
45. Tabakovic I, Grujic Z, Bejtovic Z (1983) J Heterocyclic Chem 20: 635
46. Koch D, Schafer H (1973) Angew Chem Int Ed 12: 245
47. Tabakovic I (1995) Electrochim Acta 40: 2809
48. Anderbert P, Bekolo H, Cossy J, Bouzide J (1995) J Electroanl Chem 389: 215
49. Steckhan E (1987) Topics in Current Chemistry 142: 1
50. Tabakovic I, Lacan M, Damoni Sh (1976) Electrochim Acta 21: 621
51. Tabakovic I, Grujic Z (1982) Bull Soc Chim (Belgrade) 47: 339
52. Kovalesky AC, Shine HJ (1988) J Org Chem 53: 1973
53. Shin S, Shine HJ (1992) J Org Chem 53: 2706
54. Dreher EL, Bracht J, El-Mobayed M, Hutter P, Winter W, Rieker A (1982) Chem Ber 115: 288
55. Grujic Z, Tabakovic I, Sanicanin Z (1989) Croat Chem Acta 62: 5453
56. Crljenak S, Tabakovic I, Jeremic D, Gaon I (1983) Acta Chem Scand B37: 527
57. Batusic M, Tabakovic I, Crljenak S (1981) Croat Chem Acta 54: 397
58. Gunic J, Tabakovic I, Sanicanin Z (1990) Electrochim Acta 35: 225
59. Sanicanin Z, Juric A, Tabakovic I Trinajstic N (1987) J Org Chem 52: 4053
60. Sanicanin Z, Tabakovic I (1986) Tetrahedron Lett 27: 407
61. Sanicanin Z, Tabakovic I (1988) Electrochim Acta 33: 1595

62. Miljkovic S, Tabakovic I, Sanicanin Z, Cekovic Z (1990) J Serb Chem Soc 55: 131
63. Sanicanin Z, Tabakovic I, unpublished results
64. Pandey G, Krishna A, Kumaraswamy G (1987) Tetrahedron Lett 28: 4615
65. Esker JL, Newcomb M (1993) Adv Heterocyclic Chem 58: 1
66. Tokuda M, Yamada Y, Takagi T, Suginome H (1985) Tetrahedron Lett 26: 6085
67. Tokuda M, Yamada Y, Takagi T, Suginome H (1987) Tetrahedron 43: 281
68. Tokuda M, Miyamoto T, Fujita H, Suginome H (1991) Tetrahedron 47: 747
69. Karady S, Corley EG, Abramson NL, Weinstock LM (1989) Tetrahedron Lett 30: 2191
70. Karady S, Corley EG, Abramson NL, Amato JS, Weinstock LM (1991) Tetrahedron 47: 757
71. Saito K, Hattori M, Sato T, Takahashi K (1992) Heterocycles 34: 129
72. Lacan M, Rogic V, Tabakovic I, Galijas D, Solomon T (1983) Electrochim Acta 28: 199
73. Chiba T, Sakagami H, Murata M, Okimoto M (1995) J Org Chem 60: 6764
74. Uneyama K, Kobayashi M (1994) J Org Chem 59: 3003
75. Uneyama K, Kobayashi M (1991) Tetrahedron Lett 32: 5981
76. Becker JY, Zinger B (1982) Tetrahedron 38: 1677
77. Becker JY, Zinger B (1980) Electrochim Acta 25: 791
78. Becker JY, Zinger B (1982) J Chem Soc Perkin 2: 395
79. Becker JY (1985) Isr J Chem 25: 196
80. Becker JY, Zinger B, Yatziv S (1988) J Org Chem 52: 2783
81. Becker JY, Yatziv S (1988) J Org Chem 53: 1744
82. Becker JY, Shakkour E, Sarma JARP (1990) J Chem Soc Chem Commun 1016
83. Becker JY, Shakkour E, Sarma JARP (1992) J Org Chem 57: 3716
84. Becker JY, Shakkour E (1993) Tetrahedron 49: 6285
85. Becker JY, Shakkour E (1994) Tetrahedron 50: 12773
86. Papandopouls A, Lewale B, Steckhan E, Ginzel K, Nieger M (1991) Tetrahedron 47: 563
87. Shono T, Matsumura Y, Katoh S, Takeuchi K, Sasa K, Kamada T, Shimizu R (1990) J Am Chem Soc 112: 2368
88. Abou-Elenien GM, El-Anadouli BE, Baraka RM (1991) J Chem Soc Perkin 1: 1377
89. Chen C, Shih C, Swenton JS (1986) Tetrahedron Lett 27: 1391
90. Swenton JS, Shih C, Chen C, Chou C (1990) J Org Chem 55: 2019
91. Zhang F, Dryhurst G (1991) J Org Chem 56: 7113
92. Cariou M, Carlier R, Simonet J (1983) J Chem Soc Chem Commun 876
93. Chiba T, Okimoto M (1992) J Org Chem 57: 1375
94. Schmittel M (1994) Topics in Current Chemistry 169: 183
95. Brettle R, Perkin JG (1967) J Chem Soc C: 1352
96. Lacan M, Tabakovic I (1973) Bull Soc Chim (Belgrade) 38: 297
97. Lacan M, Tabakovic I, Vukicevic M (1973) Croat Chem Acta 45: 1973
98. Schäfer H, Al Azrak (1972) Chem Ber 105: 2398
99. Torii S, Uneyama K, Onishi T, Fujita Y, Ishigiro M, Nishida T (1980) Chem I Lett 1603
100. Yoshida J, Sakaguchi K, Isoe S (1986) Tetrahedron Lett 28: 6075
101. Yoshida J, Sakaguchi K, Isoe S (1987) Tetrahedron Lett 28: 667
102. Yoshida J, Nakatani S, Isoe S (1990) Tetrahedron Lett 31: 2425
103. Shono T, Chuankamnerdkarn M, Maekawa H, Ishifune M, Kashimura S (1994) Synthesis 895
104. Nicolaou KC, Petasis NA (1984) Selenium in Natural Product Synthesis, CIS Inc, Philadelphia
105. Torii S, Uneyama K, Ono M (1980) Tetrahedron Lett 21: 2741; Torii S, Uneyama K, Ono M (1981) J Am Chem Soc 103: 4606
106. Mihailovic Mlj, Konstantinovic S, Vukicevic R (1987) Tetrahedron Lett 28: 4348
107. Konstantinovic S, Vukicevic R, Mihailovic MLj (1987) Tetrahedron Lett 28: 6511
108. (a) Vukicevic R, Konstantinovic S, Mihailovic MLj (1988) J Serb Chem Soc 53: 713; (b) Vukicevic R, Konstantinovic S, Mihailovic MLj (1991) Tetrahedron 47: 859
109. Töteberg-Kaulen S, Steckhan E (1988) Tetrahedron 44: 4389
110. Cekovic Z, Tabakovic I, unpublished results; for EGA-catalyzed reactions, see Uneyama K (1987) Topics Curr Chem 142: 167
111. Cekovic Z, Tabakovic I (1979) Bull Soc Chim (Belgrade) 44: 409
112. Miller LL, Nordblom GD, Mayeda EA (1972) J Org Chem 37: 916
113. Raju KVS, Raju PVN, Raju GGV (1990) Bull Electrochem 6: 877
114. Azzem MA, Zahran M, Hagagg E (1994) Bull Chem Soc Jpn 67: 1390

115. (a) Shono T (1984) Tetrahedron 40: 811; (b) Shono T (1988) Topics Curr Chem 148: 131; (c) Shono T (1984) in: Electroorganic Chemistry as a New Tool in Organic Synthesis, Springer-Verlag, Berlin
116. Irie K, Okita M, Wahamatu T, Ban Y (1980) Nouv J Chim 4: 275
117. (a) Moeller KD, Rothfus SL, Wong PL (1991) Tetrahedron 47: 583; (b) Moeller KD (1986) Topics Curr Chem, this issue
118. Cornille F, Slomeczynska U, Smith ML, Bensen DD, Moeller KD, Marshall GR (1995) J Am Chem Soc 117: 909
119. Chiba K, Sonoyama J, Tada M (1995) J Chem Soc Chem Commun 1381
120. Tabakovic I, Trkovnik M, Batusic M, Tabakovic K (1979) Synthesis 590
121. Berube D, Cauquis G, Pierre G, Fahmy HM (1982) Electrochim Acta 27: 281
122. Torii S, Tanaka H, Inokuchi T (1988) Topics in Current Chemistry 148: 154
123. Kitahara K, Takano Y, Nishi H (1987) Denki Kagaku 7: 1497
124. Kuhn R, Munzing W (1952) Chem Ber 85: 29
125. Tabakovic I, Trkovnik M, Galijas D (1978) J Electroanal Chem 86: 241
126. Henning N, Dassler T, Jugelt W (1981) Z Chem 22: 25
127. Pechman H, Runge P (1894) Ber 27: 323
128. Zhivich AB, Koldobskii GI, Ostrovski VA (1991) Khim Geterosikl Soedin 12: 1587
129. Lacan M, Tabakovic I, Cekovic Z (1974) Tetrahedron 30: 2911
130. Abou-Elenien G (1994) J Electroanal Chem 357: 301
131. Gollas B, Speiser B (1992) Angew Chem Int Ed 31: 332
132. Speiser B, Stahl H (1992) Tetrahedron Lett 33: 4429
133. Dunsch L, Gollas B, Neudeck A, Petr A, Speiser B, Stahk H (1994) Chem Ber 127: 2423
134. Wei X, Speiser B (1995) Electrochim Acta 40: 2477
135. Khmelnitckii LI, Novikov SS, Godovikova TI (1981) Chemistry of Furoxanes. Structure and Synthesis (in Russian) Nauka, Moscow
136. Jugelt W, Tismer M, Rank M (1983) Z Chem 23: 29
137. Petrosyan VA, Niyazymbetov ME, Ulyanova EV (1989) Izv Akad Nauk SSSR Ser Khim 7: 1683
138. Petrosyan VA, Niyazymbetov ME, Ulyanova EV (1989) Izv Akad Nauk SSSR Ser Khim 7: 1687
139. Petrosyan VA, Niyazymbetov ME, Ulyanova EV (1990) Izv Akad Nauk SSSR Ser Khim 3: 625
140. Niyazymbetov ME, Ulyanova EV, Petrosyan VA (1990) Izv Akad Nauk SSSR Ser Khim 3: 630
141. Niyazymbetov ME, Ulyanova EV, Petrosyan VA (1992) Soviet Electrochem 28: 449
142. Rolfs A, Brosig H, Liebscher J (1995) J Prakt Chem 337: 310
143. Matsuki T, Hu NX, Otsubo T, Ogura F (1988) Bull Chem Soc Jpn 61: 2117

Organic Electroreductive Coupling Reactions Using Transition Metal Complexes as Catalysts

Jean-Yves Nédélec, Jacques Périchon, Michel Troupel

Laboratoire d'Electrochimie, Catalyse et Synthèse Organique CNRS - Université Paris 12 Val de Marne; 2, rue H Dunant F-94320 Thiais, France

Table of Contents

1 Introduction 142

2 Homo-Coupling of Organic Halides 144
   2.1 Homo-Coupling of Organic Monohalides 144
   2.2 Homo-Coupling of Organic Dihalides 148

3 Cross-Coupling of Organic Halides 149
   3.1 Cross-Coupling of Aryl Halides 149
   3.2 Cross-Coupling Between Aryl- and Activated Alkyl-Halides . . . 151

4 Addition of Organic Halides to Unsaturated Groups 152
   4.1 Addition Reactions to C,C Double and Triple Bonds 152
      4.1.1 Additions to Electron-Rich C,C Double or Triple Bonds . . 153
      4.1.2 Additions to Electron-Poor Olefins 156
   4.2 Addition Reactions to Carbonyls 158

5 Synthesis of Carboxylic Acids 163
   5.1 Carboxylation of Organic Halides 163
   5.2 Carboxylation of Alkenyl and Alkynyl Compounds 164

6 Carbonylation of Organic Halides 167

7 Miscellaneous 168

8 Conclusion 169

9 Appendix 170

10 References 170

It is about 20 years since the combination of transition-metal catalysis and electroreduction was shown to be applicable to the coupling of organic molecules. This was followed by a number of fundamental investigations and basic syntheses using various nickel, cobalt, or palladium compounds which can easily be reduced in situ electrochemically to low-valent reactive intermediates. The last decade has been less characterized by reports on new catalytic systems than by the development of new synthetic applications. The aim of this review is to show that the electrochemical processes described here offer valuable advantages in organic synthesis.

1 Introduction

Modern organic chemistry intends to combine the search for new reactions with the aim of optimizing the efficiency of known and new synthetic procedures. Economical as well as environmental pressures have led chemists to attempt to increase the selectivity of the reactions while avoiding the formation of polluting by-products and using the more simple reaction conditions.

In this context, homogeneous catalysis has been increasingly used since it can offer valuable advantages, notably in terms of selectivity and efficiency. Indeed, low-valent complexes of transition metals like nickel, palladium, or cobalt can react with many functionalities, thus allowing numerous C,C-bond forming reactions.

Electrochemistry is connected to this field in many ways. For example, the synthesis of many organometallics can be carried out electrochemically by the anodic dissolution of a metal electrode in the presence of the desired ligand. Electrochemical methods can also help to investigate the reactivity of coordination compounds towards various inorganic or organic substrates and allow one to highlight the possible reaction mechanisms. These preparative and mechanistic aspects have been well investigated and recently reviewed [1, 2]. Another interesting aspect of electrochemistry is its use in organic synthesis, which will be illustrated below with many examples.

From a general point of view, reactions involving transition metal complexes can be divided into two classes.

The first class includes non-redox reactions like isomerisation, dimerisation or oligomerisation of unsaturated compounds, in which the role of the catalyst lies in governing the kinetic and the selectivity of thermodynamically feasible processes. Electrochemistry associated to transition metal catalysis has been first used for that purpose, as a convenient alternative to the usual methods to generate in situ low-valent species which are not easily prepared and/or handled [3]. These reactions are not, however, typical electrochemical syntheses since they are not faradaic; they will not be discussed in this review.

Reactions which involve both a transition metal catalyst and stoichiometric amounts of electrons either from a reductant or from an electrochemical device belong to the second class. This class can be further subdivided into two groups according to the mode of action of the catalyst. Indeed, a transition metal complex can act as electron carrier from the source of electron to the reagent. Such a process usually proceeds according to the so-called outer-sphere mechanism. We suggest that such an electron carrier be referred to as a redox mediator. Electrochemistry can be used to perform the regeneration of the mediator all over the reaction. Reactions proceeding according to an outer-sphere mechanism will not be covered in this review. The second group of faradaic and catalyzed reactions includes those reactions obeying an inner-sphere mechanism, i.e., involving the formation of various coordination compounds along the reaction pathway. Such an approach was first described about 20 years ago and already presented in reviews [4–7]. We have decided to cover results published during the last decade. We will find that, besides the fundamental aspects, many new synthetic applications have been reported.

There is a temptation in doing such a review to exaggerate the advantages of the eletrochemical method. But it would not be right to say that all reactions described here are more advantageously carried out by electrochemistry than by conventional chemical methods. We can however first point out that the electron is a cheap, readily available reagent whose redox potential can be adjusted at will. Also, electrochemistry can eventually allow us to generate unusual low-valent intermediates which are hardly or not attainable when chemical reductants are used. Thirdly, according to mechanistic investigations, there are many cases for which some intermediate of the catalytic cycle has or has not to be reduced for the reaction to proceed. This can require the tuning of the reducing power of the medium, and this may be easily done in the electrochemical approach by adjusting the working-electrode potential. All these advantages will be illustrated below by examples showing either the improvement of an existing chemical route or a new reaction.

This should convince all that electrochemistry is a real synthetic method. It is commonly said that the electrochemical method requires complex and expensive devices. However, important progress has also been made to make it simpler and not too expensive. Large numbers of electrochemical reactions are now run at constant current, thus requiring very simple electric power supplies compared to those used for controlled potential reactions. In addition, undivided cells are increasingly employed, thus avoiding the use of separators which are not very efficient when used with polar aprotic solvents, and require the use of large amounts of supporting electrolyte. With respect to this, the sacrificial anode process, which has been developed during the period covered by this review, has allowed the use of quite simple electrolytic devices, which can be more easily scaled up than divided cells. In addition the metallic ions derived from the anode increase the conductivity and can also influence the reactivity of the chemical intermediates. This device can be used for direct as well as for catalyzed reactions [8]. In the following, the utilization of the sacrificial anode method will be referred to by the indication of the nature of the anode in equations, schemes, and tables. In the absence of this, the use of a divided cell should be assumed.

As a guide for going through this review it can be helpful to consider briefly the electrochemical and chemical behavior of the most common transition metal complexes[1], i.e., nickel, palladium and cobalt, used in the various reactions collected here. Additional data can be found in [1].

Two classes of nickel catalyst precursors will appear below: one includes nickel(II) complexes leading to the corresponding nickel(0) species, the other one nickel(II) complexes leading to nickel(I) intermediates.

The first group includes nickel(II) salts ligated to either 2,2'-bipyridine (bpy), phenanthroline (phen), triphenylphosphine, or bis-diphenylphosphinoethane (dppe). In the case of nitrogen bidentate ligands (bpy or phen) the reduction at ca. -1.2 V vs SCE leads to the corresponding nickel(0) species. A stable Ni^0 species usually involves eight extra electrons, that is for example from 2 bpy or 2 dppe. However, the formation of low-ligated complexes is more convenient, because they are the more reactive as they are unstable. This emphasizes the interest of the electrochemical approach since such active species can be formed in situ and react instantaneously with the reagents.

The second group includes nickel salts ligated to a tetradentate ligand like cyclam and related compounds (CR, tet a), or salen. The electroreduction at around -1.6 V vs SCE of these compounds gives the corresponding nickel(I) intermediate which is only further reduced at quite negative potential. We will see that the two classes of nickel compounds display different behavior towards organic halides.

Cobalt catalyst precursors are cobalt(III) or cobalt(II) salts ligated to nitrogen and eventually oxygen-containing polydentate molecules like B_{12}, salen, $C_2(DO)(DOH)_{pn}$. The III/II electroreduction occurs at around 0 V vs SCE. Further reduction at ca. -1 V vs SCE corresponds to the formation of cobalt(I) complexes which are the reactive species involved in the reactions mentioned below.

Only Pd-PPh$_3$ compounds have been investigated. The electrochemical reduction of $Pd^{II}X_2(PPh_3)_2$ is a two-electron reduction occurring at ca. -1 V vs SCE and leading to the corresponding Pd^0 reactive species.

2 Homo-Coupling of Organic Halides

2.1 Homo-Coupling of Organic Monohalides

The reductive dimerisation of organic halides is, to our knowledge, the first reported reaction involving both a faradaic electroreduction and a catalysis by

---

[1] Names and fomulas of ligands are indicated in the Appendix.

transition metal complexes. It was indeed in 1976 that Jennings reported the electrodimerisation of alkyl bromides in DMF by electrolysis between two metallic electrodes (Ni, Al, or Cu) and in the presence of iron- or nickel-acetylacetonate and triphenylphosphine [9].

A few other examples involving alkyl halides have since been reported using various nickel or cobalt complexes [10–19]. Some results published during the last decade are given in Table 1. Though yields can be high, the synthetic interest of these reactions is however limited since most of these reductive dimerisations can be conducted either chemically or electrochemically without requiring any catalyst [20]. Actually, some of these studies were aimed at investigating the electrochemical behavior of catalyst precursors and the reaction mechanisms.

Biaryl synthesis from aryl halides is a more interesting reaction due to the value of these molecules and their difficult access by chemical methods. The first electrosyntheses were simultaneously done in 1979–80 by three groups [21–23] who used $NiCl_2PPh_3$ (1–20%) as catalyst precursor in the presence of excess PPh_3. Later, several groups investigated the use of bidentate phosphines like dppe associated with nickel in the synthesis of various biaryls, and notably 2,2'-bipyridine and of 2,2'-biquinoline from respectively 2-chloropyridine and 2-chloroquinoline [24]. More recently new nickel complexes with 1,2-bis(di-2-alkyl-phosphino)benzene have been studied from both fundamental and synthetic points of view [25]. They have been applied to the coupling of aryl halides.

Nickel associated with 2,2'-bipyridine has also been studied and the low valent complexes obtained by electroreduction are very efficient catalysts in the dimerisation of aryl compounds [26]. In addition, they have been used in combination with the sacrificial anode process which allows the running of such

Table 1. Ni- or Co-catalyzed electroreductive coupling of alkyl halides

No	Substrate	Metal/Ligand	Solvent	Product	Yield %	Ref.
1	ethyl bromide	Ni/phen	DMF	butane		[10]
2	RBr	Ni/bpy	DMF or NMP	R–R		[11]
	R = n-C_6H_{13}				85	
	n-C_5H_{11}				89	
	$MeOC_2H_4$				60	
	$MeOCOC_5H_{10}$				65	
	$Cl(CH_2)_4$				38	
3	allyl chloride	Co/bpy	water + micelles	1,5-hexadiene		[12]
4	$PhCHCl_2$	Ni/salen	DMF	$PhCH_2CH_2Ph$ + $PhCH=CHPh$		[15]
5	$PhCHCl_2$	Co/salen	DMF	$PhCH_2CH_2Ph$ + $PhCH=CHPh$		[16, 17]
6	X–$(CH_2)_n$–Cl X = Br, I; n = 2–6	Ni/salen		Cl–$(CH_2)2n$–Cl	80–90	[18, 19]

reactions in undivided cells under very simple experimental conditions (Eq. 1):

$$2ArX + 2e \xrightarrow[\text{DMF, Mg anode}]{\text{NiBr}_2\text{bpy (7\%) + 2bpy}} Ar\text{-}Ar + 2 X^-$$

Ar = Ph, 3-MeC$_6$H$_4$, 70-90% (1)
4-MeCOC$_6$H$_4$, Naphthyl...
X = I, Br, Cl

In the case of orthosubstituted aryl halides the corresponding arylnickel intermediate Ar$_2$Nibpy is stable and should be oxidized to lead to the coupling product and regenerate the nickel(II) compound [27].

The electrosynthesis of biaryls can also be performed with PdCl$_2$(PPh$_3$)$_2$ as catalyst precursor. The starting reagent can be either a bromo- or iodo-aryl compound (Eq. 2) [28], or an aryltriflate (Eq. 3) [29, 30], thus allowing the use of phenol compounds as starting reagents in place of the corresponding halocompounds, but at a higher temperature:

$$2ArX + 2e \xrightarrow[\text{DMF, RT}]{\text{PdCl}_2(\text{PPh}_3)_2 \text{ (7\%)}} Ar\text{-}Ar + 2 X^-$$

Ar = Ph, 4-t-BuC$_6$H$_4$, 50-98% (2)
4-Me$_2$NC$_6$H$_4$,
4-MeOC$_6$H$_4$...
X = I, Br

$$2ArOTf + 2e \xrightarrow[\text{DMF, 90 °C}]{\text{PdCl}_2(\text{PPh}_3)_2 \text{ (10\%)}} Ar\text{-}Ar + 2 TfO^-$$

Ar = Ph, 4-CNC$_6$H$_4$, 50-70% (3)
4-CF$_3$C$_6$H$_4$,
4-ClC$_6$H$_4$...

These nickel- or palladium-catalyzed homocoupling reactions can also be conducted in the presence of zinc as reducing agent, leading usually to comparable results as shown recently in the palladium-catalyzed dimerisation of aryltriflates [30].

The electrochemical analytical methods have afforded much information on the reaction mechanisms. The general case deals with the formation of a zerovalent nickel or palladium complex at the cathode when the ligand is mono- or bi-dentate like PPh$_3$, dppe, or bpy. The first step of the catalytic cycle is the oxidative addition of the organic halide to the zerovalent-complex leading to a σ-aryl-nickel or a σ-aryl-palladium intermediate. The rate constants have been determined for these reactions which are of first-order with respect to each reagent (ArX and Ni^0 or Pd^0). It comes out that the low-ligated compounds like Ni^{0}bpy [31], Ni^{0}dppe [32], or Pd^0(PPh$_3$)$_2$ [33] are much more reactive than the corresponding fully coordinated complexes. The reactivity of the Ni^0 or Pd^0 species also depends on the nature and the concentration of the halide ions associated with the starting complex or added to the medium [31, 33].

Several routes have been proposed starting from the σ-arylnickel and the σ-arylpalladium intermediate.

A detailed investigation of the Ni-dppe catalyzed formation of biphenyl from bromobenzene has resulted in the proposal of a catalytic cycle (Scheme 1) in which the σ-aryl-nickel intermediate is first reduced into the corresponding ArNi^I then transformed into a diaryl-nickel(III) complex. This complex then undergoes a reductive elimination leading to the product and Ni^I, followed by the regeneration of the Ni^0 system [32, 34].

Two catalytic cycles having the σ-aryl-nickel intermediate in common should be taken into account for the nickel-bpy system, as illustrated in Scheme 2. In the first cycle (left-hand side), similar to the previous one with dppe as ligand, Ni^{I}X which is formed in the reductive elimination step disproportionates into Ni^0 and Ni^{II} followed by a further reduction of Ni^{II} [35]. An alternative mechanism has also been proposed in which the product results from the

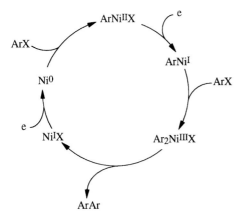

Scheme 1. Electroreductive dimerisation of aryl halides catalyzed by Ni-dppe complexes

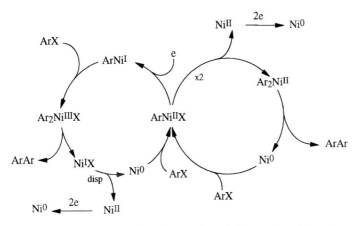

Scheme 2. Electroreductive dimerisation of aryl halides catalyzed by Ni-bpy complexes

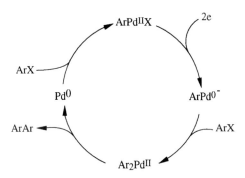

Scheme 3. Electroreductive dimerisation of aryl halides catalyzed by Pd-PPh$_3$ complexes

metathesis of the σ-aryl-nickel^{II} complex and regeneration of Ni^{II} (Scheme 2, right-hand side) [35, 36].

The Pd-PPh$_3$ system (Scheme 3) is characterized by a two-electron reduction step of the σ-aryl-palladium intermediate [37], as also proposed previously for aryl-nickel complexes ligated to PPh$_3$ [23, 38]. The formation of the biaryl proceeds by reductive elimination from the diarylpalladium and regeneration of Pd^0.

Thus, there are several different mechanisms involved in the nickel- or palladium-catalyzed dimerisation of aryl halides. Especially in the case of the Ni-bpy system, as indicated in Scheme 2, one of the two routes can become the more favorable by changing the experimental conditions, notably the cathode potential.

2.2 Homo-Coupling of Organic Dihalides

The electroreductive coupling of organic dihalides is an interesting approach to synthesize polymers.

The preparation of poly-ynes from diiodoacetylene has recently been carried out in the presence of Ni/dppe complexes (Eq. 4) [39]:

$$I-\!\!\equiv\!\!-I + 2n\,e \xrightarrow[\text{DMF}]{\text{NiI}_2\text{dppe (5\%)}} +\!\!\equiv\!\!\!\!-)_n + 2n\,I^- \quad (4)$$

More attention has been devoted to aromatic and heteroaromatic substrates since first reported in 1983 [40]. The results are shown in Table 2 [25, 41–51]. All these reactions were run with nickel complexes associated with a phosphane or bpy ligand. Depending on the experimental conditions, the polymers were either precipitated during the electrolysis or deposited as films at the surface of the electrode. The method is also convenient to prepare copolymers from a mixture of two aryl dihalides. A mechanistic investigation on the nickel-bpy catalyzed polymerisation has been reported very recently [52].

Table 2. Reductive electropolymerisation of aryl dihalides using nickel catalysts

No	Reagent	Metal/Ligand	Solvent	Polymer	Ref.
1	Cl—⟨C₆H₄⟩—Cl	Ni/dppe	DMSO	precipitate	[25]
2	Br—⟨C₆H₄⟩—⟨C₆H₄⟩—Br	Ni/dppe	DMAC	film	[41]
3	2,6-dibromonaphthalene	Ni/PPh₃	MeCN	precipitate	[42]
4	Br—⟨C₆H₄⟩—CONH—⟨C₆H₄⟩—NHCO—⟨C₆H₄⟩—Br	Ni/bpy	DMAC	precipitate	[43]
5	Br—⟨C₆H₄⟩—NHCO—⟨C₆H₄⟩—CONH—⟨C₆H₄⟩—Br	Ni/bpy	DMAC	precipitate	[43]
6	3,6-dibromocarbazole, R = alkyl, aminoalkyldisiloxane	Ni/bpy	DMAC	film or precipitate	[44] [45] [46]
7	2,5-dibromopyridine	Ni/bpy + PPh₃	MeCN	film or precipitate	[47] [48]
8	2,5-dibromofuran	Ni/bpy	MeCN	film	[49]
9	3,6-dibromocarbazole + Br—⟨C₆H₄⟩—⟨C₆H₄⟩—Br	Ni/bpy	DMAC	copolymer film or precipitate	[50] [51]

3 Cross-Coupling of Organic Halides

3.1 Cross-Coupling of Aryl Halides

Unsymmetrical biaryls, notably those having both an electron-releasing and an electron-withdrawing group, have interesting physical properties in the field of non-linear optics. Their synthesis is not easy and some attempts to prepare them from a mixture of two different aryl halides by combination of electrochemistry and transition metal catalysis with nickel or palladium complexes have been reported. The main drawback for such an approach relates to the selectivity due to the competitiveness between cross- and homo-couplings. Notably, if the two halides have similar reactivity in the oxidative addition step, a 2:1:1 statistical distribution of the three possible compounds, respectively the unsymmetrical and the two symmetrical, is obtained. If the reactivity of the two compounds is quite different, an acceptable yield can be obtained only if one adjusts the reagent concentration ratio by slowly adding the most reactive halide throughout the electrolysis.

This was done for the coupling of two aryl halides having an electron-withdrawing group on one and an electron-releasing group on the other in the presence of Ni-bpy complexes (Eq. 5) [53]. The reactions were run in an undivided cell in the presence of a magnesium anode in NMP at room temperature. Good selectivity was obtained in some cases, though not as high as desired for a preparative route:

$$R^1-\text{C}_6\text{H}_4-X + R^2-\text{C}_6\text{H}_4-Cl \xrightarrow[\text{NMP, Mg-anode}]{\text{NiBr}_2\text{bpy (4-18\%)} + 2\text{bpy, e}} R^1-\text{C}_6\text{H}_4-\text{C}_6\text{H}_4-R^2 \quad (5)$$

X = Br, Cl
R^1 = MeO, Me$_2$N, MeS, R^2 = F, CN, CF$_3$

15–70%

Intramolecular cross-coupling can also be carried out as illustrated in the ring closure of Eq. 6 catalyzed by a nickel complex (L = 1,2-bis-(diisopropyl phosphino)benzene) [25]:

$$\text{(substrate)} \xrightarrow[\text{DMSO, 65 °C}]{e, \text{NiCl}_2\text{L}} \text{(phenanthridine product)} \quad (6)$$

30%

So far a high selectivity in the electrochemical synthesis of unsymmetrical biaryls has only been obtained in two-step processes.

This was done, for example, with a Pd-PPh$_3$ complex used stoichiometrically as outlined in Eq. 7 [28, 37, 54]. The electrochemically generated Pd^0 complex first reacts with one aryl halide. The electroreduction of this σ-Pd-complex in the presence of the other aryl halide affords the unsymmetrical biaryl in good yield:

$$\text{PdCl}_2\text{L}_2 \xrightarrow{2e, \text{PPh}_3} \text{Pd}^0\text{L}_2 \xrightarrow{Ar^1X} Ar^1\text{PdXL}_2 \xrightarrow{Ar^2X, 2e} Ar^1\text{-}Ar^2 + Pd^0L_2 \quad (7)$$

L = PPh$_3$ Ar^{1}X = 4-Me$_2$NC$_6$H$_4$I   Ar^{2}X = 4-t-BuC$_6$H$_4$I,   76%

Another two-step approach was reported using Ni-bpy as catalyst. The electroreduction of an aryl halide catalyzed by a Ni-bpy complex, when carried

Table 3. Two-step Ni- then Pd-catalyzed cross-coupling of aryl halides

$$Ar^1X \xrightarrow[\text{DMF, Zn-anode}]{\text{e, Ni(BF}_4)_2\text{bpy}_3\text{ (7\%)} + 2\text{ bpy} + \text{ZnBr}_2} Ar^1ZnX \xrightarrow[\text{PdCl}_2(\text{PPh}_3)_2\text{ (0.5-2\%)}]{Ar^2X} Ar^1Ar^2$$

No	Ar^{1}X	Ar^{2}X	Yield %
1	4-ClC$_6$H$_4$CF$_3$	4-BrC$_6$H$_4$OCH$_3$	83
2	4-ClC$_6$H$_4$CF$_3$	4-BrC$_6$H$_4$CN	83
3	4-BrC$_6$H$_4$OCH$_3$	4-BrC$_6$H$_4$CN	90
4	4-BrC$_6$H$_4$OCH$_3$	4-BrC$_6$H$_4$NO$_2$	84
5	4-ClC$_6$H$_4$CO$_2$Me	4-BrC$_6$H$_4$OCH$_3$	84
6	4-BrC$_6$H$_4$NMe$_2$	4-BrC$_6$H$_4$NO$_2$	84
7	4-BrC$_6$H$_4$NMe$_2$	4-BrC$_6$H$_4$CN	83
8	2-BrC$_6$H$_4$OCH$_3$	2-BrC$_6$H$_4$CN	55

out in the presence of an excess of both bpy and Zn^{2+} and a sacrificial anode of zinc, does not lead to the symmetrical biaryl but to an arylzinc compound, which is stable in the reaction medium [55]. The addition of another aryl halide and catalytic amounts of PdCl$_2$(PPh$_3$)$_2$ to this mixture readily gives the unsymmetrical biaryl in quite good yields (Table 3) [56]. The method also applies to the coupling between aryl halide and allylchloride [55]. The great advantage of this method is its large functional tolerance.

3.2 Cross-Coupling Between Aryl- and Activated Alkyl-Halides

Aryl-acetic or -propionic acids as well as benzylketones are versatile intermediates for the synthesis of pharmaceuticals, agrochemicals, or fragrances. Many methods have already been explored, notably using electrochemistry and transition metal compounds.

It was shown a few years ago that α-haloesters can react with an aryl-nickel compound generated electrochemically in a divided cell (Eq. 8) [57]. The main drawback for this two-step method is the use of a large excess of PPh$_3$ to prevent the homocoupling reaction:

$$PhX + 2e + NiCl_2 \xrightarrow{PPh_3} PhNiX(PPh_3)_2 \xrightarrow{X'CH(R)CO_2Et} PhCH(R)CO_2Et \quad (8)$$
$$X, X' = Cl, Br \quad R = H, Me \qquad\qquad\qquad\qquad 30\text{-}85\%$$

A more efficient method combining the use of a sacrificial anode and the catalysis by nickel complexes was reported recently. Optimal reaction conditions were found to minimize the unwanted homo-couplings, by slowly adding the most reactive reagent, i.e., the activated alkyl halide, and by running the electrolyses at 60–80 °C. The method was applied to the cross-coupling between arylhalides and either α-chloroesters (Table 4) [58, 59], α-chloroketones

Table 4. Ni-bpy catalyzed electroreductive cross-coupling between aryl halides and α-chloroesters

$$\text{ArX + ClCHRCO}_2\text{Me} \xrightarrow[\text{DMF, Zn or Al anode}]{\text{e, NiBr}_2\text{bpy (5-10\%)}} \text{ArCHRCO}_2\text{Me}$$

No	ArX	R	ArCHRCO$_2$Me Yield %
1	4-FC$_6$H$_4$Br	H	75
2	4-CNC$_6$H$_4$Br	H	60
3	4-Me$_2$NC$_6$H$_4$Br	H	67
4	4-FC$_6$H$_4$Br	Me	65
5	3-CF$_3$C$_6$H$_4$Br	Me	59
6	4-CF$_3$C$_6$H$_4$Br	Me	66
7	4-CNC$_6$H$_4$Br	Me	70
8	4-MeOC$_6$H$_4$Br	Me	51
9	4-MeOC$_6$H$_4$I	Me	85
10	2-bromo, 6-methoxy-naphthalene	Me	55

Table 5. Ni-bpy catalyzed electroreductive cross-coupling between aryl halides and α-chloroketones

$$\text{ArX + ClCHR}^1\text{COR}^2 \xrightarrow[\text{DMF, Zn or Al anode}]{\text{e, NiBr}_2\text{bpy (5-10\%)}} \text{ArCHR}^1\text{COR}^2$$

No	ArX	R^1	R^2	ArCHR^{1}COR^2 Yield %
1	2-FC$_6$H$_4$Br	H	Me	56
2	3-CF$_3$C$_6$H$_4$Br	H	Me	80
3	4-MeO$_2$CC$_6$H$_4$Br	H	Me	52
4	4-MeOC$_6$H$_4$I	H	Me	65
5	4-MeOC$_6$H$_4$I	H	Ph	52
6	3-CF$_3$C$_6$H$_4$Br	H	Ph	63
7	3-CF$_3$C$_6$H$_4$Br	Me	Me	70
8	4-CNC$_6$H$_4$Br	Me	Me	70
9	4-MeOC$_6$H$_4$Br	Me	Me	53
10	3-bromopyridine	Me	Me	45

(Table 5) [59–61], or allylic or vinylic derivatives (Table 6) [60] to give the corresponding coupling products in good yields.

4 Addition of Organic Halides to Unsaturated Groups

4.1 Addition Reactions to C,C Double and Triple Bonds

Electroreductive addition reactions of organic halides to C,C double or triple bonds have been investigated since the early 1980s, and two main methods have

Table 6. Ni-bpy catalyzed electroreductive cross-coupling between aryl halides and vinylic or allylic derivatives

$$\text{ArX} + \text{RX} \xrightarrow[\text{DMF, Zn or Al anode}]{\text{e, NiBr}_2\text{bpy (5-10\%)}} \text{ArR}$$

No	ArX	RX	ArR Yield %
1	3-CF$_3$C$_6$H$_4$Br	BrCH=CH–CH$_3$ (Z/E = 50/50)	60 (Z/E = 20/80)
2	4-CNC$_6$H$_4$Br	BrCH=CH–CH$_3$ (Z/E = 50/50)	66 (Z/E = 20/80)
3	4-MeOC$_6$H$_4$Br	BrCH=CH–CH$_3$ (Z/E = 50/50)	44 (Z/E = 25/75)
4	3-CF$_3$C$_6$H$_4$Br	AcOCH$_2$CH=CH$_2$	56
5	4-CF$_3$C$_6$H$_4$Cl	AcOCH$_2$CH=CH$_2$	38
6	4-MeOC$_6$H$_4$Br	AcOCH$_2$CH=CH$_2$	52
7	4-MeOC$_6$H$_4$Br	AcOCH$_2$CH=CHCH$_3$	63[a,b]
8	4-MeOC$_6$H$_4$I	CH$_3$CH(Cl)CH=CH$_2$	56[a,c]

[a] ArCH=CHCH$_2$CH$_3$ + ArCH(CH$_3$)CH=CH$_2$
[b] 86/12
[c] 96/4

been proposed, both involving radical additions as the key steps. One approach uses vitamin B$_{12}$ or related cobalt complexes as mediators in an indirect electrochemical reduction, sometimes in combination with photolysis. This work has been mainly done in Scheffold's group [62, 63]. The other approach uses nickel associated with macrocyclic tetradentate ligands which can stabilise Ni^I species, as described by Gosden and Pletcher [64].

The studies done during the last decade cover various aspects:
– investigation of various Ni- or Co-organometallic compounds with regard to their redox properties and their reactivity according to the ligand associated with the metal;
– re-investigation of previously described reactions in order to improve their efficiency, notably in the case of Ni-Cyclam complexes;
– generation of carbanion equivalents via Ni^0 complexes in the presence of Michael acceptors.

The reactions reviewed here can be divided into two classes: those involving electron-rich alkenes or alkynes and which are of the radical type, and those involving electron-poor olefins (Michael acceptors) and which are of the radical or carbanion type.

4.1.1 Additions to Electron-Rich C,C Double or Triple Bonds

Most of the results recently reported in this field relate to free-radical cyclisations (Eq. 9) which are useful processes to make mainly five-membered rings via the 5-*exo*-trig mode, with eventually a given stereochemistry [65]:

(9)

X = C, O, N   major

Table 7. Co- or Ni-catalyzed electroreductive annulation reactions

No	Substrate	Catalyst	Product Yield %	Conditions Reaction	Ref.
1	R = H, Et, C_5H_{11}, $SiMe_3$; n = 1, 2; X = O, N-CO_2E	Cobaloxime (5%)	54 - 80%	MeOH 55–60 °C Zn-anode	[68]
2[a]	R^1 = H, Me Ph; R^2 = H, Me	Co-L[b] 5 or 40%	42 - 96%	acetonitrile	[70]
3	R^1 = H, R^2 = R^3 = C_6H_5; R^1, R^2 = -$(CH_2)_n$- R^3 = H	Ni(cyclam) (20 %)	50 - 86%	DMF	[71]
4	R = H, Ts, Allyl, benzyl; R^1, R^2 = H, Me	Ni(CR) (20%)	11 - 67%	DMF or acetonitrile	[72]
5[c]		Ni(CR) (20%)	7 - 33%	DMF or acetonitrile	[72]
6	R^1, R^2 = H, Me; E = CO_2R	Ni(tet a) (20%)	32 - 81%	DMF	[73]

Table 7. Continued

No	Substrate	Catalyst	Product Yield %	Conditions Reaction	Ref.
7	(benzene with X and A-CH=CH₂ substituents) A = O, NH X = I, Br	a) Ni(tet a) (20%) b) Ni(cyclam) (10%)	(bicyclic product with A) a or b : 60-90%	a) DMF b) DMF Mg-anode	[71] [73] [75]
8	Ph—≡—(CH₂)₃-CH₂-X	Ni(salen) (10 %)	(cyclopentane with Ph, H substituents) 84%	DMF	[78]

[a] saturated product obtained when conducted in with Co-L (5%) and RSH (25eq)
[b] L = C_2(DO)DOH)pn;
[c] R is identical to R in entry 4

The electrochemical processes involving cobalt complexes have already been thoroughly investigated [62, 66]. Additional results have been reported with cobalt complexes different from vitamin B_{12} and also with nickel ligated to tetradentated macrocyclic ligands. Both series of complexes lead to radicals.

The indirect cyclisation of bromoacetals via cobaloxime(I) complexes was first reported in 1985 [67]. At that time the reactions were conducted in a divided cell in the presence of a base (40% aqeous NaOH) and about 50% of chloropyridine cobaloxime(III) as catalyst precursor. It was recently found that the amount of catalyst can be reduced to 5% (turnover of ca. 50) and that the base is no longer necessary when the reactions are conducted in an undivided cell in the presence of a zinc anode [68, 69]. The method has now been applied with cobaloxime or Co[C_2(DO)(DOH)$_{pn}$] to a variety of ethylenic and acetylenic compounds to prepare fused bicyclic derivatives (Table 7, entry 1). The cyclic product can be either saturated or unsaturated depending on the amount of catalyst used, the cathode potential, and the presence of a hydrogen donor, e.g., RSH (Table 7, entry 2). The electrochemical method was found with some model reactions to be more selective and more efficient than the chemical route using Zn as reductant [70].

Nickel compounds can also be used as radical generators in electroreductive processes, according to previous work on Ni-cyclam and related complexes [64]. Three research groups have reinvestigated the process and extended its use to free-radical annulation reactions.

Ni-cyclam, Ni(CR), or Ni(tet a) can be used efficiently as catalyst in DMF, and in the presence of NH_4ClO_4 as proton source [71–74]. Ni^I species generated electrochemically react rapidly with organic halides to generate alkyl, alkenyl, or aryl radicals which add intramolecularly to a double or triple bond, then leading to cyclopentanoids (Table 7, entries 3–7a).

J.-Y. Nédélec et al.

These reactions can also be conducted in an undivided cell, as shown in the case of aryl halides bearing an unsaturated chain (Table 7, entry 7b). Cyclam and other tetradentate donor ligands associated with nickel were used, magnesium being the most convenient anode in these reactions [75–77].

Ni-salen is a good alternative to nickel-cyclam to generate Ni^I intermediates. Indeed the cyclisation of 6-phenyl-1-halo-pent-5-yne occurs very efficiently in the presence of 10% of catalyst, as compared to the direct electroreduction of the reagent (Table 7, entry 8) [78].

We can also mention that one example of electroreductive cyclization of N-alkenyl-2-bromoanilines has recently been reported using Pd-PPh$_3$ as catalyst [79].

4.1.2 Additions to Electron-Poor Olefins

Two types of intermediates, i.e., radicals or carbanions or their organometallic equivalents, can be used to perform addition reactions to Michael acceptors. The free-radical route has already been investigated with nickel or cobalt complexes as catalysts [62–64]. These studies have been reinvestigated recently with the aim of improving the turn-over of the catalyst and/or using easily prepared cheap complexes.

Interest in the vitamin B_{12} catalyzed light assisted electrolysis directed to the preparation of fine chemicals has been well illustrated in the synthesis of some natural products. One typical example is the synthesis of the "Queen substance" of *Apis mellifera* which involves two efficient electrochemical C,C-bond-forming steps by addition to conjugated double and triple bonds (Scheme 4) [63].

Other examples include the synthesis of California Red Scale Pheromone [80], prostaglandin PGF2α [81], House Mouse Pheromone [63], and Jasmonates [82].

The method was also applied to the synthesis of unusual aminoacids (Table 8; Y = NHAc) [83].

In suitable cases the intermediate radical formed in the addition step can show diastereofacial selectivity in the subsequent protonation step. This was

Table 8. Vitamin B_{12} catalyzed alkylation or acylation of activated olefins

R-X + CH$_2$=C(Y)COOMe $\xrightarrow{\text{e, B}_{12} \text{(4%) -light, DMF}}$ R-CH$_2$-C(Y)(H)COOMe + X$^-$

No	R	X	Y	Yield (%)
1	CH$_3$(CH$_2$)$_3$	Br	NHCOCH$_3$	70
2	CH$_3$(CH$_2$)$_3$	Br	H	75
3	CH$_3$CO	CH$_3$COO	NHCOCH$_3$	67
4	CH$_3$CO	CH$_3$COO	H	70
5	CH$_3$CO$_2$(CH$_2$)$_3$	Br	NHCOCH$_3$	72

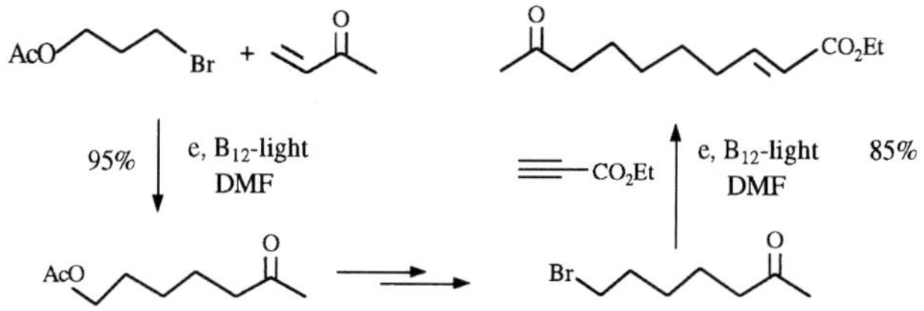
AcO
Br
95%
e, B_{12}-light
DMF
CO_2Et
e, B_{12}-light
DMF
85%
AcO
Br

Scheme 4. Synthesis of the "Queen substance" of *Apis mellifera*

Table 9. Electroreductive alkylation of activated olefins using Ni (tet a)$^{2+}$ as catalyst

R-Br + R^1-CH=C(R^2)W $\xrightarrow[\text{DMF+NH}_4\text{ClO}_4]{\text{e, Ni (tet a)}^{2+}}$ 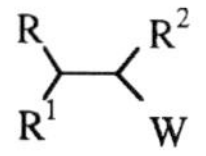
R
R^2
R^1
W

No	RBr	R^1, R^2	W	Yield (%)
1	BuBr	H, H	CO_2CH_2Ph	53
2		H, H	CO_2H	32
3	$Ph(CH_2)_3Br$	$-(CH_2)_4-$	CO	14
4		Me, H	CO_2Me	13
5		H, Me	CO_2Me	53
6		H, H	CO_2Me	57
7		H, H	CN	72
8		H, H	COMe	37
9	$Ph(CH_2)_2CHMeBr$	H, H	CO_2Me	60

observed in the vitamin B_{12}-catalyzed addition involving *t*-BuBr and diethyl mesaconate (Eq. 10) [84]. The chemical method (B_{12}/Zn) seems however to be more efficient, with no difference in the stereoselectivity which is only affected by the addition of amines:

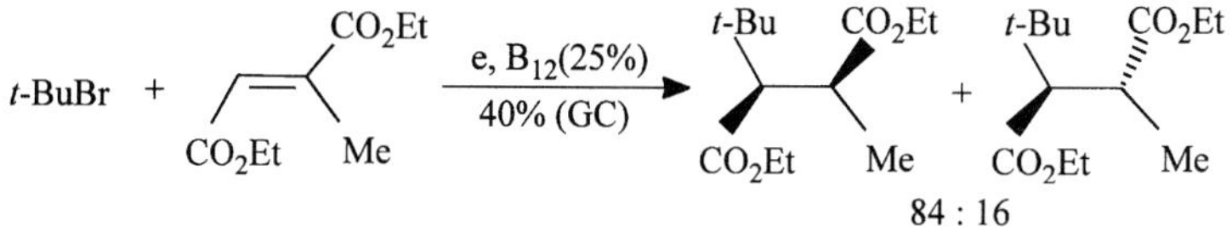
CO_2Et
t-BuBr +
CO_2Et
Me
e, B_{12}(25%)
40% (GC)
t-Bu
CO_2Et
CO_2Et
Me
+
t-Bu
CO_2Et
CO_2Et
Me

84 : 16 (10)

Nickel-cyclam and related complexes can also be used though previous reports indicated that the turnover of Ni/(tet a) in acetonitrile is low [85]. The process has now been reinvestigated to show that Ni/(tet a) can been used in catalytic conditions (2%) in DMF containing NH_4ClO_4 as proton source to perform the alkylation of unsaturated esters, ketones, or nitriles (Table 9) [86]. Yields are good if the terminal carbon of the double bond is not substituted (R^1 = H).

Nucleophilic additions can also be carried out using nickel^{II} complexes which lead to Ni^0 by cathodic reduction. Ni(0) complexes, notably Ni^0bpy, react

Table 10. Electroreductive arylation of activated olefins using Ni-pyridine as catalyst

$$\text{Ar-Br} + \text{R}^1\text{-CH=C(R}^2\text{)-W} \xrightarrow[\substack{\text{DMF-Pyridine (9:1)} \\ \text{Al or Fe- anode}}]{\text{e, NiBr}_2(5\text{-}10\%)} \underset{\text{R}^1}{\overset{\text{Ar}}{\diagdown}}\!\!\!\!\!\!\!\!\!\!\underset{\text{W}}{\overset{\text{R}^2}{\diagup}}$$

No	Ar	R^1, R^2	W	Yield (%)
1	C_6H_5	H, H	CO_2Et	63
2	$p\text{-CNC}_6H_4$	H, H	CO_2Et	48
3	2-naphthyl	H, H	CO_2Et	50
4	$p\text{-MeOC}_6H_4$	CO_2Et, H	CO_2Et	46
5	$p\text{-MeCOC}_6H_4$	CO_2Et, H	CO_2Et	31
6	2-naphthyl	H, Me	CO_2Et	27
7	2-naphthyl	Me, H	CO_2Et	20
8	2-naphthyl	H, H	CN	61
9	2-naphthyl	$-(CH_2)_4-$	CO	20

rapidly with alkyl as well as alkenyl and aryl halides. In the presence of activated olefins the organonickel intermediate formed by oxidative addition does not however lead to the corresponding addition product. A less associated ligand like pyridine has been found to be more convenient, also allowing the activated olefin to coordinate to the metal. Thus efficient addition reactions of aryl groups were performed in very simple and mild conditions. The reactions were conducted at 60 °C in an undivided cell using a sacrificial anode of aluminum or iron. Yields of 20–63% were obtained in the arylation reactions, which compare advantageously with the chemical method using aryl-copper or cuprates. The reactions are regioselective since no 1,2-addition occurred. Representative examples are given in Table 10 [87].

Intramolecular addition can also be performed using either the radical or the carbanion pathway [25, 72, 73, 88]. These reactions are presented in Table 11. The synthesis of bicyclic ketones has previously been carried out in the presence of vitamin B_{12} (5%) in good to high yields [62]. Other Ni- or Co-complexes have now been reported to give the same cyclisation reaction, though with a relatively low turnover. Among the complexes studied Ni(cyclam)(ClO$_4$)$_2$, Ni(CR)(ClO$_4$)$_2$ and Co(bpp)OAc were reported to be the most efficient ones [88].

Intramolecular arylation of unsaturated amides leading to lactams can be obtained via the radical or carbanionic pathway. The latter (Table 11, entry 6) leads to a mixture of γ- and δ-lactams in a ca. 3:1 ratio, while the radical route gives selectively the γ-lactam (Table 11, entry 5), though in moderate yield.

4.2 Addition Reactions to Carbonyls

Two main types of processes which belong to basic synthetic reactions have been investigated recently: the allylation of carbonyl compounds and the Reformatsky reaction.

Table 11. Co- or Ni-catalyzed intramolecular additions to Michaël acceptors

No	Substrate	Catalyst	Product Yield (%)	Reaction Conditions	Ref.
1	cyclohexanone with $(CH_2)_n$-Br at α-position, n = 4, 5	Co- or Ni complex (30%)	bicyclic ketone $(CH_2)_n$, 40 - 70%	DMF	[88]
2	cyclohexanone with $(CH_2)_n$-Br, n = 4, 5	Co- or Ni complex (30%)	spirocyclic ketone $(CH_2)_n$, 47 - 65%	DMF	[88]
3	cyclohexanone with Br and allyl-E substituents	Ni/(tet a)	bicyclic product with E, 86%	DMF	[73]
4	acrylamide with Br, OMe, N-Me	Ni/(tet a) (20%)	oxindole with OMe, N-Me, 23%	DMF	[72]
5	acrylamide with Br, N-Me (ortho)	Ni/PPh$_3$	oxindole (N-Me) 40% + dihydroquinolinone (N-Me) 13%	DMSO	[25]

Direct electrochemical allylation can be performed with some carbonyl compounds. Transition metal catalyzed couplings are however more efficient, notably in reactions involving ketones.

Two methods have been proposed. One involves $PdCl_2(PPh_3)_2$ as catalyst along with a zinc salt ($ZnCl_2$) for the allylation of carbonyl compounds using allylic acetate as reagent to give the corresponding alcohol in good yields (Table 12 and Scheme 5) [89]. With conjugated carbonyls, only 1,2 addition occurred.

Of the two regio-isomers which can be formed in this reaction, the branched isomer is predominant (Scheme 5).

A mechanistic study of the reaction has highlighted the key role of $ZnCl_2$ [90]. It is the most readily reduced species, and Zn^0 might be the reducing agent of the first formed π-allyl palladium intermediate leading to allylzinc compounds as illustrated in Scheme 6 for allylacetate.

The alternative method uses a nickel complex as catalyst (Table 13) [91, 92]. This reaction, which was carried out in a undivided cell, was found to be efficient

Table 12. Pd-catalyzed allylation of carbonyl compounds

$$\text{R}\diagup\!\!\diagdown\text{OAc} + \text{R}^1\text{R}^2\text{C=O} \xrightarrow[\text{DMF}]{\text{Pd}^{II}(5\%),\,\text{ZnCl}_2,\,e} \text{R}\diagup\!\!\diagdown\!\overset{\text{R}^1\,\text{R}^2}{\diagdown}\text{OH}$$

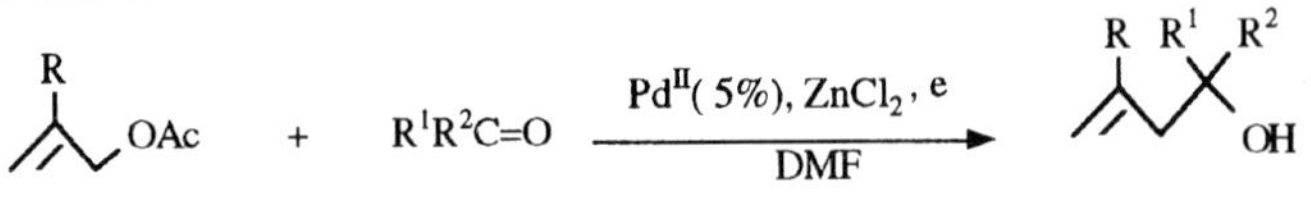

No	R	R^1, R^2	Yield (%)
1	H	C$_6$H$_5$, H	77
2	H	4-MeO-C$_6$H$_4$, H	84
3	H	4-ClC$_6$H$_4$	68
4	H	–(CH$_2$)$_5$–	40
5	CH$_3$	C$_6$H$_5$, H	62

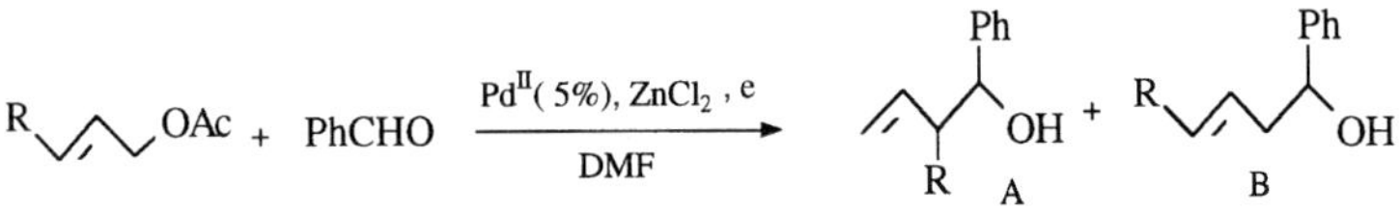

$$\text{R}\diagup\!\!\diagdown\!\!\diagup\text{OAc} + \text{PhCHO} \xrightarrow[\text{DMF}]{\text{Pd}^{II}(5\%),\,\text{ZnCl}_2,\,e} \underset{\text{A}}{\overset{\text{Ph}}{\text{R}\diagup\!\!\diagdown\!\!\diagup\text{OH}}} + \underset{\text{B}}{\overset{\text{Ph}}{\text{R}\diagup\!\!\diagdown\!\!\diagup\text{OH}}}$$

R = Me: 56% (A/B = 91 : 9)

R = Ph: 71% (A/B = 83 : 17)

Scheme 5. Regioselectivity in the Pd-catalyzed allylation of carbonyl compounds in the presence of Zn salts

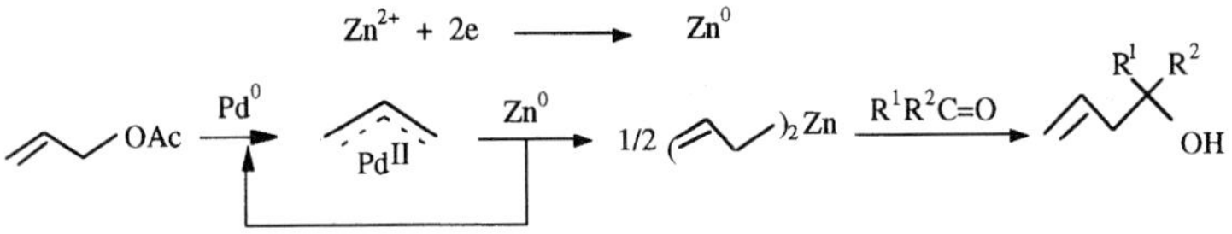

$$\text{Zn}^{2+} + 2e \longrightarrow \text{Zn}^0$$

$$\diagup\!\!\diagdown\text{OAc} \xrightarrow{\text{Pd}^0} \overset{\diagup\!\!\diagdown}{\text{Pd}^{II}} \xrightarrow{\text{Zn}^0} 1/2\,(\diagup\!\!\diagdown)_2\text{Zn} \xrightarrow{\text{R}^1\text{R}^2\text{C=O}} \diagup\!\!\diagdown\!\overset{\text{R}^1\,\text{R}^2}{\diagdown}\text{OH}$$

Scheme 6. Reaction mechanism of the Pd-catalyzed electrochemical allylation of carbonyl compounds in the presence of zinc salts

Table 13. Ni-catalyzed electrochemical allylation of carbonyl compounds

$$\text{R}\diagup\!\!\diagdown\text{X} + \text{R}_1\text{R}_2\text{C=O} \xrightarrow[\text{DMF}]{\text{NiBr}_2\text{bpy},\,e,\,\text{Zn anode}} \text{R}\diagup\!\!\diagdown\!\overset{\text{R}^1\,\text{R}^2}{\diagdown}\text{OH}$$

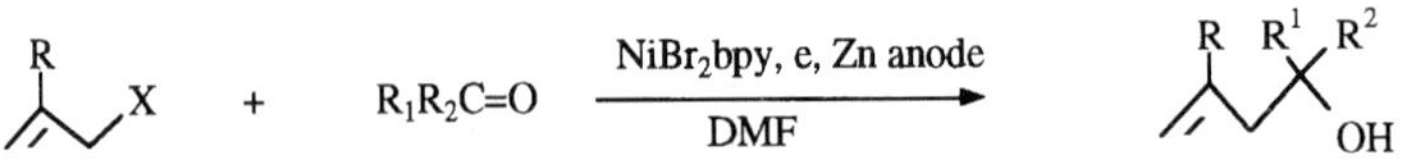

No	X	R	R^1, R^2	Yield (%)
1	Cl	Me	Ph, H	85
2	Cl	Me	CH$_3$(CH$_2$)$_5$, H	70
3	Cl	Me	Ph, Ph	86
4	Cl	Me	t-Bu, Me	60
5	Cl	Me	CH$_2$=CH, Me	43
6	OAc	H	Ph, H	80
7	OAc	H	–(CH$_2$)$_5$–	83

only in the presence of a sacrificial zinc anode. Aromatic and aliphatic aldehydes or ketones give good yields of the corresponding homoallylic alcohols. Lower yields are however obtained from hindered ketones like di-isopropylketone. With α-β-unsaturated carbonyls the 1,2-addition was exclusively observed. Allylic transposition was also observed with appropriate starting molecules as in the case of the Pd/Zn system. The authors have suggested a transmellation process between the allyl-nickel compound, first formed by oxidative addition of the allyl compound to the Ni^0 complex, and the anodically generated zinc(II) species [92, 93].

Both the palladium- and the nickel-catalysis enables the use of allylic acetate as starting reagent. In the two approaches a zinc compound is evoked as key intermediate, though its formation has been demonstrated only indirectly. In the two methods allylic transposition is observed. The authors have then concluded that these electrochemical allylation reactions closely parallel the chemical allylation reactions involving allylzinc intermediates.

The Reformatsky reaction can also be performed electrochemically either directly or using a mediator. Ni-catalysis has proven to be an efficient way to prepare β-hydroxy ester or nitrile from the corresponding α-chlorocompounds (Table 14) [94]. Here again the first step is the oxidative addition of the cathodically generated Ni^0bpy to the halocompound. The nature of the sacrificial anode also plays a crucial role in this reaction, though the formation of an organozinc intermediate has not been fully demonstrated.

An interesting application of the method is the reaction involving methyl chlorodifluoroacetate (Table 15) [95], the activation of which cannot be performed chemically. The reactions were best conducted in CH_2Cl_2-DMF (90:10) as solvent instead of DMF. Here also a transmetallation has been invoked, on the basis of ^{19}F NMR analysis of the reaction mixture.

The coupling of methyl dichloroacetate with acetophenone or cyclohexanone leads to high yields (75–80%) of epoxides under similar reaction

Table 14. Ni-catalyzed electrochemical Reformatsky reaction

$$ClCH(R)CO_2Me + R_1R_2C=O \xrightarrow[DMF]{NiBr_2bpy, e, Zn\ anode} \underset{R}{\underset{|}{R^2}}\overset{R^1}{\underset{}{\diagup}}\overset{OH}{\underset{}{\diagdown}}CH\text{-}CO_2Me$$

No	R	R^1, R^2	Yield (%)
1	H	Et, Et	77
2	H	$-(CH_2)_5-$	86
3	H	Ph, Ph	77
4	Me	n-Pr, Me	64
5	Me	Ph, H	80
6	Me	$-CH=CH(CH_2)_3-$	50

Table 15. Electrosynthesis of 2,2-difluoro-3-hydroxyesters using Ni-complex as catalyst

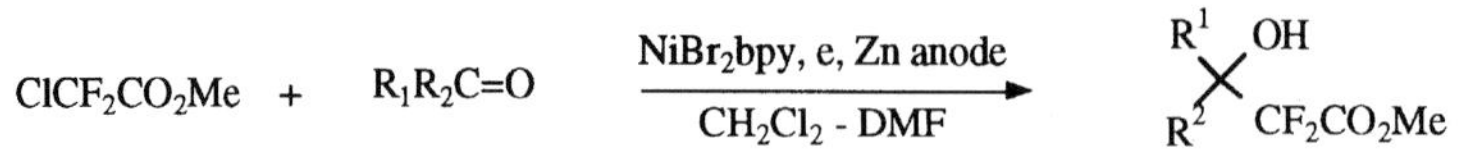

$$ClCF_2CO_2Me + R_1R_2C=O \xrightarrow[CH_2Cl_2 - DMF]{NiBr_2bpy, e, Zn\ anode} \underset{R^2}{\overset{R^1}{>}}\!\!\!\underset{CF_2CO_2Me}{\overset{OH}{<}}$$

No	R^1, R^2	Yield (%)
1	Ph, Me	72
2	$-(CH_2)_5-$	63
3	Ph, H	70
4	$CH_3(CH_2)_2$, H	77
5	5-Me-furyl, H	45

conditions (Eq. 11) [94]:

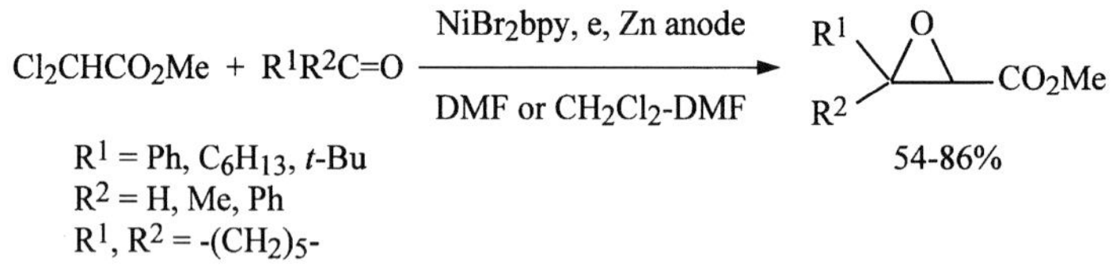

$$Cl_2CHCO_2Me + R^1R^2C=O \xrightarrow[DMF\ or\ CH_2Cl_2-DMF]{NiBr_2bpy,\ e,\ Zn\ anode} \underset{R^2}{\overset{R^1}{>}}\!\!\triangle\!-CO_2Me \quad (11)$$

R^1 = Ph, C_6H_{13}, t-Bu
R^2 = H, Me, Ph
$R^1, R^2 = -(CH_2)_5-$
54–86%

The process can be extended to the coupling reaction of chloroacetonitrile or α-chloropropionitrile (Eq. 12) [94].

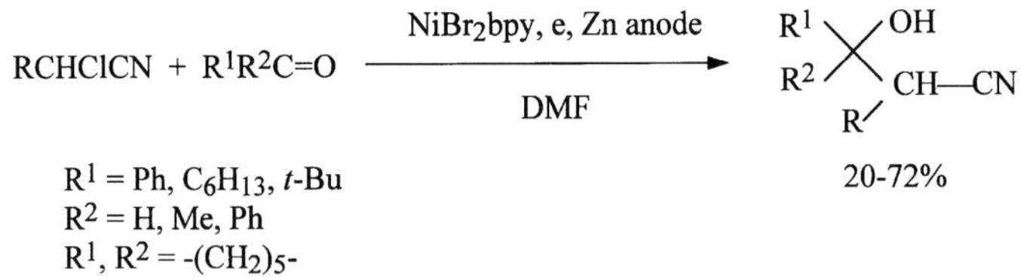

$$RCHClCN + R^1R^2C=O \xrightarrow[DMF]{NiBr_2bpy,\ e,\ Zn\ anode} \underset{R^2}{\overset{R^1}{>}}\!\!\!\underset{R'}{\overset{OH}{<}}\!\!CH-CN \quad (12)$$

R^1 = Ph, C_6H_{13}, t-Bu
R^2 = H, Me, Ph
$R^1, R^2 = -(CH_2)_5-$
20–72%

Intramolecular addition to the carbonyl group has not been much investigated. The only example is the ring enlargement of α-bromomethyl-cycloalkanones conducted in the presence of $Co[C_2(DO)(DOH)_{pn}]Cl_2$ as catalyst at 55–60 °C [69, 96]. The reaction is thought to go through a radical pathway in keeping with the general behavior of alkyl cobalt intermediates. The turnover of the catalyst and the yields have been improved by the combined use of the cobalt catalysis and a sacrificial anode of zinc (Eq. 13) [69]:

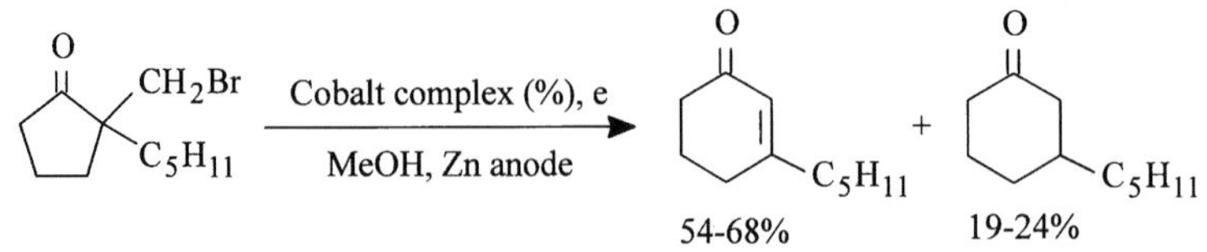

$$\underset{}{\overset{O}{\underset{}{\bigcirc}}}\!\!\!\overset{CH_2Br}{\underset{C_5H_{11}}{<}} \xrightarrow[MeOH,\ Zn\ anode]{Cobalt\ complex\ (\%),\ e} \underset{54-68\%}{\overset{O}{\bigcirc}\!-\!C_5H_{11}} + \underset{19-24\%}{\overset{O}{\bigcirc}\!-\!C_5H_{11}} \quad (13)$$

5 Synthesis of Carboxylic Acids

5.1 Carboxylation of Organic Halides

In the early 1980s it was shown that the electroreduction of aryl halides catalyzed by Ni-PPh$_3$ [97] or Ni-dppe [98] and in the presence of CO_2 mainly leads to the arylcarboxylate instead of the biaryl. An electroanalytical study of the Ni-dppe system has resulted in the proposal of a catalytic cycle [99, 100]. In this mechanism CO_2 is involved in a reaction with the aryl-nickel(I) formed by electroreduction of the σ-aryl-nickel(II) as indicated in Scheme 1.

The electrocarboxylation of aryl iodides or bromides can also be catalyzed by the Pd-PPh$_3$ system (Eq. 14) [101].

$$RC_6H_4X + CO_2 + 2e \xrightarrow[DMF]{PdCl_2(PPh_3)_2\ (7\%)\ +\ 2\ PPh_3} RC_6H_4CO_2^- + 2\ X^- \quad (14)$$

R = H, 4-t-Bu, 4-MeO
X = I, Br
50–90%

It has been proposed that in this reaction CO_2 reacts as an electrophile with [ArPd^0(PPh$_3$)$_2$]$^-$ formed by reduction of the σ-aryl-palladium(II) [102]. Aryl chlorides react too slowly with Pd^0 to enable an efficient carboxylation reaction. On the other hand aryl triflate and aryl bromide have similar reactivity. The synthesis of aryl carboxylic acids can then be obtained from phenols via the formation of the corresponding aryltriflate (Eq. 15) [29, 30]:

$$ArOTf + CO_2 + 2e \xrightarrow[DMF,\ 90\ °C]{PdCl_2(PPh_3)_2\ (10\%)} ArCO_2^- + TfO^- \quad (15)$$

Ar = R-C$_6$H$_4$
with R = 4-Cl, 4-CF$_3$, 4-CO$_2$Et, 4-F, 4-CH$_3$
Ar = 1- or 2-naphthyl
52–95%

Non-steroidal anti-inflammatory α-arylpropionic acids were also prepared from the corresponding benzylic chlorides and CO_2 using as catalyst Ni-dppe or Ni-dppp in the presence of COD (Table 16) [103]. The use of the catalyst in this reaction is not absolutely required but its use limits the homocoupling reaction which would be the main process at high concentration of the benzylic halide and low pressure of CO_2 [104].

The catalyst is not necessary either for the electrocarboxylation of aryl halides or various benzylic compounds when conducted in undivided cells and in the presence of a sacrificial anode of aluminum [105] or magnesium [8, 106]. Nevertheless both methods, i.e., catalysis and sacrificial anode, can be eventually associated in order to perform the electrocarboxylation of organic halides having functional groups which are not compatible with a direct electroreductive process.

Table 16. Ni(dpp)(COD) catalyzed electrocarboxylation of benzylic chlorides

$$\text{ArCH(Me)Cl} + \text{CO}_2 \xrightarrow[\text{THF + HMPA}]{\substack{\text{NiCl}_2\text{dpp (10\%)}\\ \text{COD (10\%), e}}} \text{ArCH(Me)CO}_2\text{H}$$

No	Product		Yield (%)
1	PhCH(Me)CO$_2$H		89
2	[MeO-naphthyl-CH(Me)CO$_2$H]	naproxen	81
3	[iBu-C$_6$H$_4$-CH(Me)CO$_2$H]	ibuprofen	80
4	[PhO-C$_6$H$_4$-CH(Me)CO$_2$H]	fenoprofen	76

5.2 Carboxylation of Alkenyl and Alkynyl Compounds

The direct carboxylation of unsaturated compounds has already been reported [107, 108]. The process is however limited to reducible alkenes like aryl olefins or activated olefins. On the other hand, only a few examples of direct electrocarboxylation of alkynes have been reported [109].

The indirect electroreductive coupling between alkynes and CO_2 leading to substituted acrylic acids via a hydrocarboxylation process has been recently investigated [110–114]. The reactions were conducted in an undivided cell, in DMF containing catalytic amounts of a nickel complex, either Ni(bpy)$_3$(BF$_4$)$_2$ or NiBr$_2$dme + 2PMDTA depending on the substrate, in the presence of a magnesium sacrificial anode and a carbon cathode, at normal or 5 atmosphere pressure of CO_2. The reactions were run at 20–80 °C; a high temperature favored the carboxylation of alkynes having electron-releasing groups while reactions involving compounds with electron-withdrawing groups were best conducted at room temperature. Typical results obtained for terminal and disubstituted alkynes with Ni(bpy)$_3$(BF$_4$)$_2$ as catalyst are given in Table 17 [114]. The reaction occurs with a good regioselectivity in the introduction of CO_2 at the 2-position of terminal alkynes. It is also stereoselective, occurring mainly as a *cis*-addition. Only the monocarboxylation is obtained unless an electron-withdrawing group is present (Table 17, entry 5). Yields, based on a 30-85% consumption of the alkyne, are good.

Table 17. Electrochemical carboxylation of alkynes in the presence of $Ni(bipy)_3(BF_4)_2$

$$R_1-\!\!\!\equiv\!\!\!-R_2 + CO_2 \xrightarrow[\text{2) hydrolysis}]{\text{1) Ni(bipy)}_3(BF_4)_2(10\%),\ e,\ DMF,\ Mg\text{-anode}} \underset{\mathbf{a}}{\overset{R_1\quad R_2}{\underset{CO_2H\ \ H}{>\!\!=\!\!<}}} + \underset{\mathbf{b}}{\overset{R_1\quad R_2}{\underset{H\ \ CO_2H}{>\!\!=\!\!<}}}$$

No	R^1	R^2	Yields (%)	a:b
1	$n\text{-}C_6H_{13}$	H	65	90:10
2	$n\text{-}C_3H_7$	$n\text{-}C_3H_7$	93	
3	C_6H_5	H	55	71:29
4	C_6H_5	CH_3	72	38:62
5	C_6H_5	CO_2Et	75[a]	

[a] Di- and monocarboxylation occurred in a 74/26 ratio.

The method can be favorably compared to the chemical method developed by Hoberg, who has used stoichiometric amounts of a zerovalent nickel complex, mostly the air-sensitive $Ni(COD)_2$ [115, 116]. Instead, the electrochemical method uses a readily available starting Ni(II) complex.

The advantage of using a sacrificial anode has been clearly pointed out. Magnesium was found to be the most convenient, the oxidation of which produces Mg^{2+} ions which can enter the catalytic cycle to cleave the nickelacycle intermediate and liberate Ni for further catalytic cycles (Scheme 7). Such a mechanism has been substantiated on the basis of the formation of the nickelacycle and its characterization by cyclic voltammetry. In the absence of Mg^{2+} (reactions conducted in a divided cell in the presence of ammonium ions) the nickelacycle does not transform and the reaction stops when all the starting nickel compound has been reacted. Upon addition of $MgBr_2$ to an electrochemically prepared solution of the nickelacycle, Ni(II) is recovered [114].

Diynes have already been used for building polycyclic compounds; in the presence of CO_2 and a stoichiometric amount of Ni(0) bicyclic pyrones were obtained [117]. With the electrocarboxylation method, linear or cyclic monocarboxylic acids were obtained as main products from non-conjugated diynes depending on the ligand associated to Ni [118, 119]. Thus ring-fomation occurred with the Ni-bipyridine complex at normal pressure of CO_2; on the other hand, in the presence of PMDTA as ligand and with a 5 atmosphere pressure of CO_2, linear adducts were mainly formed as illustrated in Eq. (16):

	yield	a	b	
L = bpy	35%	59	4.5	(16)
L = PMDTA	30%	9	45	

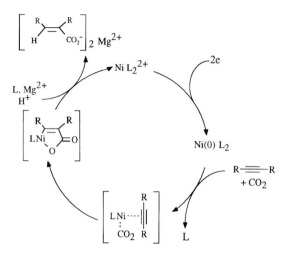

Scheme 7. Electrochemical carboxylation of alkynes catalyzed by nickel complexes

Table 18. Electrocarboxyaltion of 1,3-diynes in the presence of Ni-PMDTA catalyst

No	R^1	R^2	Yield (%)
1	n-C_5H_{11}	n-C_5H_{11}	a: 58; b: 25
2	Ph	Ph	a: 40; b̄: 0
3	$MeOCH_2$	$MeOCH_2$	a: 60; b̄: 1
4	Ph	n-Bu	a: 37; b̄: 16; c: 3
5	$MeOCH_2$	Ph	a: 50; b̄: 0; c̄: 11

Various diynes were thus converted into the corresponding linear monocarboxylic acid. Diynes having both a terminal and an internal triple bond exclusively gave the product with CO_2 attached to the terminal bond mostly at the 2-position.

The electrocarboxylation of 1,3-diynes in the presence of Ni-PMDTA as catalyst led to (E)-2-vinylidene-3-yne carboxylic acids regio- and stereo-selectively [119, 120]. Yields referred to a 70–100% conversion are good (Table 18).

In this process, double bonds were found to be less reactive than triple bonds. Thus norbornene or styrene were hydrocarboxylated in low yields (10–40%) [121]. In unconjugated as well as conjugated ene-ynes, only the alkyne moiety was carboxylated with regio- and stereoselectivity similar to that observed for alkynes [122].

1,2-Dienes have also been investigated in this process [123]. Monocarboxylic acids were obtained in good yield, with introduction of CO_2 mainly at

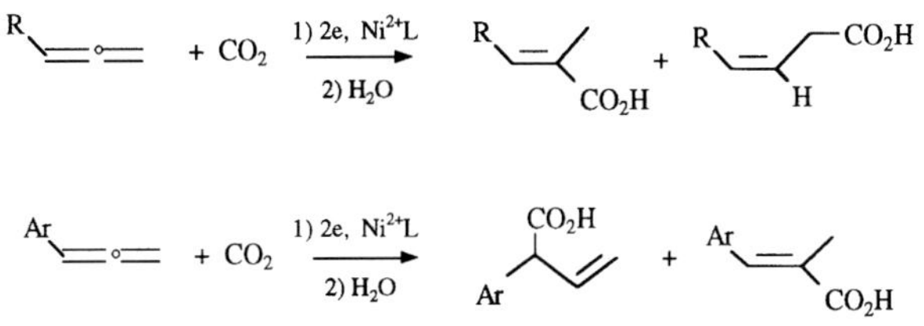

Scheme 8. Ni-catalyzed electrocarboxylation of 1,2-dienes

C_2 for R = alkyl, cycloalkyl and at C_3 for R = aryl (Scheme 8, L = PMDTA; P_{CO_2} = 5 atmospheres).

6 Carbonylation of Organic Halides

Carbon monoxide and low valent transition metals are known to give various quite well described complexes. However, due to the strong coordination to CO, these metal carbonyl compounds are not very reactive towards carbon-halogen bonds. Thus the carbonylation of organic halides remains a difficult reaction since the presence of CO leads to the deactivation of the catalytic system. Various attempts to overcome this drawback have however been reported.

Thus it has been shown that some metal-carbonyl compounds can be activated by electrochemical reduction generating reactive anionic species. Without going into details, it is worth pointing out that the synthesis of aldehydes can be obtained by electrolyzing a stoichiometric mixture of alkyl halides and ironpentacarbonyls (Eq. 17) [124, 125]:

$$RX + Fe(CO)_5 \xrightarrow[DMF \text{ or } AN]{1)\ e\quad 2)\ H^+} RCHO \qquad (17)$$

R = n-C_5H_{11}, C_2H_5, $PhCH_2$
X = I, Br
30-70%

Alternatively, CO_2 can be used as source of CO. Indeed, it is well known that low-valent transition metal complexes can catalyze the chemical or electrochemical reduction of CO_2 into CO. This approach was used to generate the mixed nickel complex $Ni^0 bpy(CO)_2$ by the electrochemical reduction of $Nibpy^{2+}$ in NMP or DMF in the presence of CO_2. The reduced complex can react with alkyl, benzyl, and allylhalides to give the symmetrical ketone along with the regeneration of $Nibpy^{2+}$. A two-step method alternating electroreduction and chemical coupling leading to the ketone has thus been set up (Scheme 9) [126, 127].

It has been shown very recently that carbonylation reactions can also be performed by electrolyzing under CO a solution of organic halide containing

$$\text{Ni(bpy)}_2^{2+} \xrightarrow[\text{Mg-anode}]{\text{e, CO}_2 \atop \text{DMF or NMP}} \text{Ni}^0\text{bpy(CO)}_2 + \text{CO}_3^{2-}$$

$$\text{Ni}^0\text{bpy(CO)}_2 \xrightarrow{\text{RX}} \text{RCOR} + \text{Ni(bpy)}_2^{2+}$$

RX = PhCH$_2$Cl, C$_6$H$_{13}$Br
CH$_3$CH=CHCH$_2$Cl

yields : 70 - 85%

Scheme 9. Ni-catalyzed electrosynthesis of ketones using CO$_2$

Table 19. Electrosynthesis of ketones from organic halides and carbon monoxide

$$2\,\text{RX} + \text{CO} + 2e \xrightarrow[\text{DMF or NMP}]{\text{FeCl}_2\,(5\%);\,\text{bpy}\,(5\%) \atop \text{stainless steel anode}} \text{RCOR} + 2\,\text{X}^-$$

No	RX	Yield (%)
1	Ph-CH$_2$Cl	80
2	2-CH$_3$-C$_6$H$_4$-CH$_2$Br	70
3	4-CH$_3$-C$_6$H$_4$-CH$_2$Br	62
4	2-Cl-C$_6$H$_4$-CH$_2$Cl	75
5	CH$_3$-CH=CH-CH$_2$Cl	65
6	CH$_3$-(CH$_2$)$_5$Br	62
7	4-CF$_3$-C$_6$H$_4$Br	50

a catalytic amount of bpy in an undivided cell fitted with a stainless steel anode to give symmetrical ketones in high yields (Table 19) [128]. The anodic reaction affords nickel(II)bpy and iron(II)bpy complexes which are reduced at the cathode, then leading to the synthesis of ketones. The reactive species is certainly a nickel complex ligated to both bpy and CO, but iron complexes are likely to play a synergic role.

7 Miscellaneous

Apart from the electrocarbonylation reactions of organic halides described in Sect. 6, other Ni-catalyzed reactions leading to ketones have been reported. Thus the electroreductive coupling between acylchlorides and alkyl halides, catalyzed by NiBr$_2$bpy leads to unsymmetrical ketones [129]. Recently acylchlorides have been converted to symmetrical ketones in an undivided cell fitted with a nickel or stainless steel anode. In this reaction the active metallic species are formed by electroreduction of Ni^{2+} derived from the anode, in the absence of added ligand (Eq. 18) [130].

$$2\ RCOCl + 2\ e \xrightarrow[\text{stainless steel anode}]{\text{MeCN}} RCOR + 2\ Cl^- + CO \quad (18)$$

R = $PhCH_2$, Ph, 3- or 4-FC_6H_4, thienyl-CH_2 45-80%

Arylzinc compounds electrochemically generated in the presence of Ni-bpy and zinc salts, already mentioned in Sect. 3.1, can also react with $(CF_3CO)_2O$ to give the corresponding aryltrifluoromethylketones [55].

Very few transition-metal catalyzed electroreductive carbon-heteroatom bond formations have been described. The electrochemical silylation of allylic acetates was carried out in the presence of Pd-PPh_3 [131]. The electrosynthesis of arylthioethers from thiophenol and aryl halides [132] and the coupling of bromobenzene with dichlorophenylphosphine [133] were performed with Ni-bpy as catalyst.

8 Conclusion

Many interesting synthetic applications combining electroreduction and transition metal catalysis have thus appeared during the last decade. Not all but many of them can really compete with the conventional chemical methods as new efficient synthetic routes. During the same period, direct electroreductive couplings have also been much investigated, notably in connection with the sacrificial anode process. We have mentioned in the review that some reactions previously described as catalyzed processes, e.g., the carboxylation of organic halides or the homocoupling of alkyl halides, can now be carried out by direct electroreductive coupling in the presence of a sacrificial anode. The transition-metal catalysis associated with the electrochemistry remains however very useful, notably when the selectivity is concerned. The most recent investigations have even highlighted for some processes a synergy between a transition metal catalyst and the metallic ions derived from the anode. Since in many cases the electrochemical devices have become more simple, there is no doubt that direct and indirect electrochemical methods will become more and more attractive.

9 Appendix: names and formulas of ligands or complexes

bpy 2,2′-bipyridine

COD 1,5-cyclooctadiene

dme 1,2-dimethoxyethane

PMDTA N,N,N′,N″,N″-pentamethyl diethylenetriamine

dppe 1,2-bis(diphenylphosphino)-ethane

dppp 1,3-bis(diphenylphosphino)propane

phen 1,10-phenanthroline

M (salen)

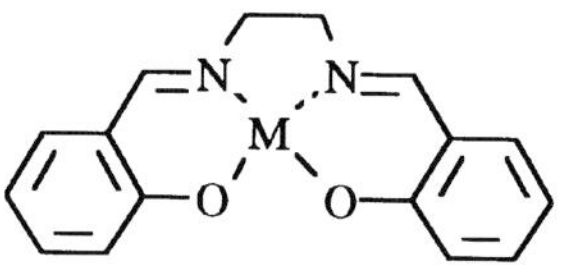

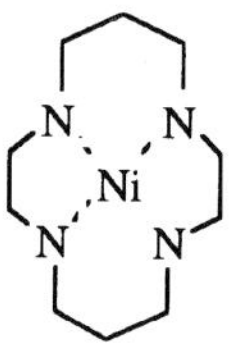

M (bpp)

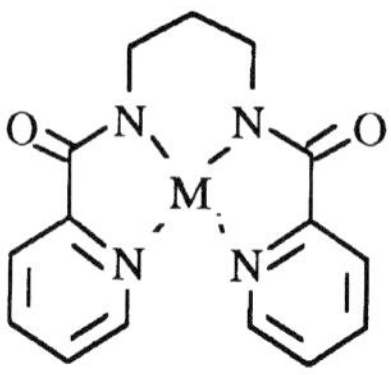

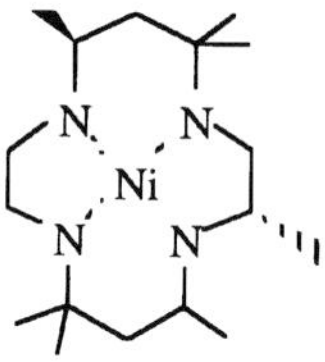

cobaloxime

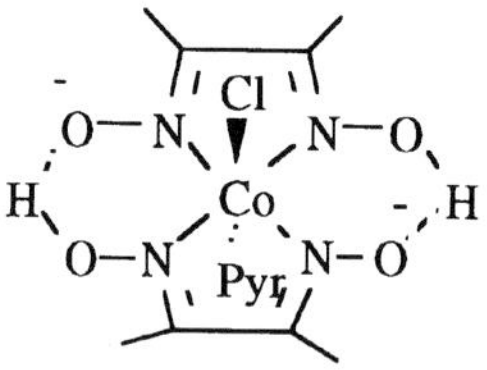

Co[C₂(DO)(DOH)$_{pn}$]

Ni (cyclam)

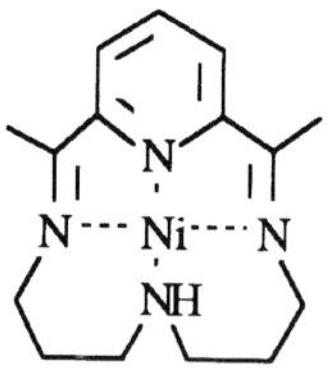

Ni (tet a)

Ni (CR)

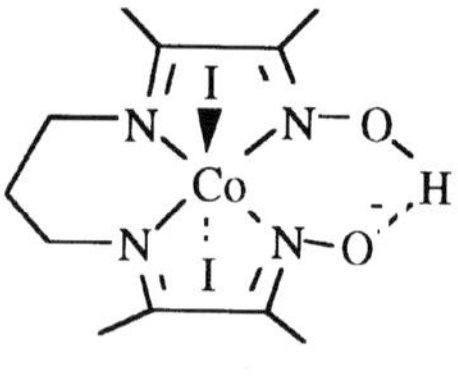

10 References

1. Walder L (1991) Organoelemental and coordination compounds. In: Lund H, Baizer MM (eds) Organic electrochemistry, 3rd edn. Dekker, New York Basel Hong Kong, p 809
2. Chakravorti MC, Subrahmanyam GVB (1994) Coord Chem Rev 135/136: 65
3. Lehmkuhl H (1973) Synthesis 377

4. Troupel M (1986) Annali di Chimica 76: 151
5. Torii S (1986) Synthesis 873
6. Steckhan E (1987) Top Curr Chem 142: 1
7. Efimov ON, Strelets VV (1990) Coord Chem Rev 99: 15
8. Chaussard J, Folest JC, Nédélec JY, Périchon J, Sibille S, Troupel M (1990) Synthesis 369
9. Jennings PW, Pillsbury DG, Hall JL, Brice VT (1976) J Org Chem 41: 719
10. Smith WH, Kuo YM (1985) J Electroanal Chem 188: 189
11. Mabrouk S, Pellegrini S, Folest JC, Rollin Y, Périchon J (1986) J Organomet Chem 301: 391
12. Kamau GN, Rusling JF (1988) J Electroanal Chem 240: 217
13. Rusling JF (1991) Acc Chem Res 24: 75
14. Fry AJ, Sirisoma UN, Lee AS (1993) Tetrahedron Lett 34: 809
15. Fry AJ, Fry PF (1993) J Org Chem 58: 3496
16. Fry AJ, Sirisoma UN (1993) J Org Chem 58: 4919
17. Fry AJ, Sirisoma N, Singh AH, Uglioloro A, Lee A, Kaufman S, Phanijphand T (1995) In: Torii S (ed) Novel Trends in Electroorganic Synthesis, Kodansha, Tokyo, p 83
18. Mubarak MS, Peters DG (1995) J Electroanal Chem 388: 195
19. Peters DG, Dahm CE, Bhattacharya D, Butler AL, Mubark MS (1995) In: Torii S (ed) Novel Trends in Electroorganic Synthesis, Kodansha, Tokyo, p 67
20. Nédélec JY, Folest JC, Périchon J (1989) J Chem Res (s) 394
21. Troupel M, Rollin Y, Sibille S, Fauvarque JF, Périchon J (1980) J Organomet Chem 202: 435
22. Mori M, Hashimoto Y, Ban Y (1980) Tetrahedron Lett 21: 631
23. Schiavon G, Bontempelli G, Corain B (1981) J Chem Soc, Dalton Trans 1074
24. Sock O, Troupel M, Périchon J, Chevrot C, Jutand A (1985) J Electroanal Chem 183: 237
25. Fox MA, Chandler DA, Lee C (1991) J Org Chem 56: 3246
26. Rollin Y, Troupel M, Tuck DG, Périchon J (1986) J Org Chem 303:131
27. Meyer G, Rollin Y, Périchon J (1987) J Organomet Chem 333: 263
28. Torii S, Tanaka H, Morisaki K (1985) Tetrahedron Lett 26:1655
29. Jutand A, Négri S, Mosleh A (1992) J Chem Soc, Chem Commun 1729
30. Jutand A, Mosleh A, Negri S (1995) In: Torii S (ed) Novel Trends in Electroorganic Synthesis, Kodansha, Tokyo, p 217
31. Troupel M, Rollin Y, Sock O, Meyer G, Périchon J (1986) N J Chem 10: 593
32. Amatore C, Jutand A (1988) Organomet 7: 2203
33. Amatore C, Azzabi M, Jutand A (1991) J Amer Chem Soc 113: 8375
34. Amatore C, Jutand A, Mottier L (1991) J Electroanal Chem 306:125
35. Durandetti M, Devaud M, Périchon J N J Chem, in press
36. Yamamoto T, Wakabayashi S, Osakada K (1992) J Organomet Chem 428: 223
37. Amatore C (1995) In: Torii S (ed) Novel Trends in Electroorganic Synthesis, Kodansha, Tokyo, p 227
38. Troupel M, Rollin Y, Sibille S, Fauvarque JF, Périchon J (1980) J Chem Res (s) 26
39. Kijima M, Sakai Y, Shirakawa H (1994) Chem Lett 2011
40. Fauvarque JF, Petit MA, Pfluger F, Jutand A, Chevrot C, Troupel M (1983) Makromol Chem 4: 455
41. Aboulkassim A, Chevrot C (1993) Polymer 34: 401
42. Tomat R, Zecchin S, Schiavon G, Zotti G (1988) J Electroanal Chem 252: 215
43. Chevrot C, Benazzi T, Barj M (1995) Polymer 36: 631
44. Siove A, Ades D, N'Gbilo E, Chevrot C (1990) Synth Met 38:331
45. Helary G, Chevrot C, Sauvet G, Siove A (1991) Polym Bull 26: 131
46. Aboulkassim A, Faïd K, Siove A (1993) Macromol Chem 194: 29
47. Schiavon G, Zotti G, Bontempelli G, Lo Coco F (1988) Synth Met 25: 365
48. Schiavon G, Zotti G, Bontempelli G, Lo Coco F (1988) J Electroanal Chem 242: 131
49. Zotti G, Schiavon G, Comisso N, Berlin A, Pagani G (1990) Synth Met 36:337
50. Aboulkassim A, Faïd K, Chevrot C (1994) J Appl Polym Sci 52: 1569
51. Faïd K, Adès D, Siove A, Chevrot C (1994) Synth Met 63: 89
52. Siove A, Aboulkassim A, Faïd K, Adès D (1995) Polym Int 37: 171
53. Meyer G, Troupel M, Périchon J (1990) J Organomet Chem 393: 137
54. Amatore C, Carré E, Jutand A, Tanaka H, Qinghua, R, Torii S submitted
55. Sibille S, Ratovelomanana V, Périchon J (1992) J Chem Soc, Chem Comm 283
56. Sibille S, Ratovelomanana V, Nédélec JY, Périchon J (1993) Synlett 425
57. Folest JC, Périchon J, Fauvarque JF, Jutand A (1988) J Organomet Chem 342: 259

58. Conan A, Sibille S, d'Incan E, Périchon J (1990) J Chem Soc, Chem Comm 48
59. Durandetti M, Sibille S, Nédélec JY, Périchon J (1995) In: Torii S (ed) Novel Trends in Electroorganic Synthesis, Kodansha, Tokyo, p 209
60. Durandetti M, Nédélec JY, Périchon J (1996) J Org Chem, 61: 1748
61. Durandetti M, Sibille S, Nédélec JY, Périchon J (1994) Synth Commun 24: 145
62. Scheffold R, Abrecht S, Orlinski R, Ruf HR, Stamouli P, Tinembart O, Walder L, Weymeuth C (1987) Pure Appl Chem 59: 363
63. Scheffold R (1991) In: Electroorganic Synthesis, Festschrift for M.M. Baizer (eds) Little RD, Weinberg NL Dekker, New York, p 317
64. Gosden G, Pletcher D (1980) J Organomet Chem 186: 401
65. Beckwith LJ, Kawrence T, Serelis AK (1980) J Chem Soc, Chem Comm 484
66. Walder L, Orlinski R (1987) Organomet 6: 1606
67. Torii S, Inokuchi T, Yukawa T (1985) J Org Chem 50: 5875
68. Inokuchi T, Kawafuchi H, Aoki K, Yoshida A, Torii S (1994) Bull Chem Soc Jpn 67: 595
69. Inokuchi T (1995) In: Torii S (ed) Novel Trends in Electroorganic Synthesis, Kodansha, Tokyo, p 223
70. Giese B, Erdmann P, Göbel T, Springer R (1992) Tetrahedron Lett 33: 4545
71. Ozaki S, Matsushita H, Ohmori H (1992) J Chem Soc, Chem Comm 1120
72. Ozaki S, Matsushita H, Ohmori H (1993) J Chem Soc Perkin Trans 1: 2339
73. Ozaki S, Horiguchi I, Matsushita H, Ohmori H (1994) Tetrahedron Lett 35: 725
74. Ozaki S, Mitoh S, Urano Y, Ohmori H (1995) In: Torii S (ed) Novel Trends in Electroorganic Synthesis, Kodansha, Tokyo, p 185
75. Olivero S, Duñach E (1994) Synlett 531
76. Olivero S, Clinet JC, Duñach E (1995) Tetrahedron Lett 36: 4429
77. Clinet JC, Duñach E (1995) J Organomet Chem in press
78. Mubarak MS, Peters DG (1992) J Electroanal Chem 332: 127
79. Tanaka H, Ren O, Torii S (1995) In: Torii S (ed) Novel Trends in Electroorganic Synthesis, Kodansha, Tokyo, p 195
80. Auer L, Weymuth C, Scheffold R (1993) Helv Chim Acta 76: 810
81. Busato S, Tinembart O, Zhang ZD, Scheffold R (1990) Tetrahedron 46: 3155
82. Busato S, Scheffold R (1994) Helv Chim Acta 77: 92
83. Orlinski R, Stankiewicz T (1988) Tetrahedron Lett 29: 1601
84. Erdmann P, Schäfer J, Springer R, Zeitz HG, Giese B (1992) Helv Chim Acta 75: 639
85. Healy KP, Pletcher D (1978) J Organomet Chem 161: 109
86. Ozaki S, Matsushita H, Ohmori H (1993) J Chem Soc Perkin Trans 1 649
87. Condon-Gueugnot S, Léonel E, Nédélec JY, Périchon J (1995) J Org Chem 60: 7684
88. Ozaki S, Nakanishi T, Sugiyama M, Miyamoto C, Ohmori H (1991) Chem Pharm Bull 39: 31
89. Qiu W, Wang Z (1989) J Chem Soc, Chem Comm 356
90. Zhang P, Zhang W, Zhang T, Wang Z, Zhou W (1991) J Chem Soc, Chem Comm 491
91. Sibille S, d'Incan E, Leport L, Massebiau MC, Périchon J (1987) Tetrahedron Lett 28: 55
92. Durandetti S, Sibille S, Périchon J (1989) J Org Chem 54: 2198
93. Sibille S, Nédélec JY, Périchon J (1991) In: Electroorganic Synthesis, Festschrift for MM. Baizer (eds) Little RD, Weinberg NL Dekker, New York, p 361
94. Conan A, Sibille S, Périchon J (1990) J Org Chem 56: 2018
95. Mcharek S, Sibille S, Nédélec JY, Périchon J (1991) J Organomet Chem 401: 211
96. Inokuchi T, Tsuji M, Kawafuchi H, Torii S (1991) J Org Chem 56: 5945
97. Troupel M, Rollin Y, Périchon J, Fauvarque JF (1981) N J Chem 5: 621
98. Fauvarque JF, Chevrot C, Jutand A, François M, Périchon J (1984) J Organomet Chem 264: 273
99. Amatore C, Jutand A (1991) J Am Chem Soc 113: 2819
100. Amatore C, Jutand A (1991) J Electroanal Chem 306: 141
101. Torii S, Tanaka H, Hamatani T, Morisaki K, Jutand A, Pflüger F, Fauvarque JF (1986) Chem Lett 169
102. Amatore C, Jutand A, Khalil F, Nielsen MF (1992) J Am Chem Soc 114: 7076
103. Fauvarque JF, Jutand A, François M (1986) Nouv J Chim 10: 119
104. Fauvarque JF, Jutand A, François M (1988) J Appl Electrochem 18: 109
105. Silvestri G, Gambino S, Filardo G, Gulotta A (1984) Angew Chem, Int Ed Engl 23: 979
106. Gal J, Folest JC, Troupel M, Moingeon MO, Chaussard J (1995) N J Chem 19: 401
107. Baizer MM (1984) Tetrahedron 40: 944 and references therein

108. Shono T (1984) Electroorganic Chemistry as a New Tool in Organic Synthesis, Springer-Verlag, Berlin
109. Wawzonek S, Wearring D (1959) J Am Chem Soc 81: 2067
110. Duñach E, Périchon J (1988) J Organomet Chem 352: 239
111. Labbé E, Duñach E, Périchon J (1988) J Organomet Chem 353: C51
112. Duñach E, Dérien S, Périchon J (1989) J Organomet Chem 364: C33
113. Duñach E, Périchon J (1990) Synlett 1: 143
114. Dérien S, Duñach E, Périchon J (1991) J Am Chem Soc 113: 8447
115. Hoberg H, Schaefer D, Burkhart G, Krüger C, Romao M (1984) J Organomet Chem 266:203
116. Hoberg H, Bärhausen D (1989) J Organomet Chem 379: C7
117. Tsuda T, Morikawa S, Hasegawa N, Saegusa T (1990) J Org Chem 55: 2878
118. Dérien S, Duñach E, Périchon J (1990) J Organomet Chem 385: C43
119. Dérien S, Clinet JC, Duñach E, Périchon J (1993) J Org Chem 58: 2578
120. Dérien S, Clinet JC, Duñach E, Périchon J (1991) J Chem Soc, Chem Comm 549
121. Dérien S, Clinet JC, Duñach E, Périchon J (1992) Tetrahedron 48: 5235
122. Dérien S, Clinet JC, Duñach E, Périchon J (1992) J Organomet Chem 424: 213
123. Dérien S, Clinet JC, Duñach E, Périchon J (1990) Synlett 1: 361
124. Vanhoye D, Bedioui F, Mortreux A, Petit F (1988) Tetrahedron Lett 29: 6441
125. Yoshida K, Kunugita EI, Kobayashi M, Amano SI (1989) Tetrahedron Lett 30: 6371
126. Garnier L, Rollin Y, Périchon J (1989) New J Chem 13: 53
127. Garnier L, Rollin Y, Périchon J (1989) J Organomet Chem 367: 347
128. Oçafrain M, Devaud M, Troupel M, Périchon J (1995) J Chem Soc, Chem Comm, 2331
129. Marzouk H, Rollin Y, Folest JC, Nédélec JY, Périchon J (1989) J Organomet Chem 369: C47
130. Folest JC, Pereira-Martins E, Troupel M, Périchon J (1993) Tetrahedron Lett 34: 7571
131. Torii S, Tanaka H, Katoh T, Morisaki K (1984) Tetrahedron Lett 25: 3207
132. Meyer G, Troupel M (1988) J Organomet Chem 354: 249
133. Budnikova YH, Yusupov AM, Kargin YM (1995) In: Torii S (ed) Novel Trends in Electroorganic Synthesis, Kodansha, Tokyo, p 187

Author Index Volumes 151–185

Author Index Vols. 26–50 see Vol. 50
Author Index Vols. 51–100 see Vol. 100
Author Index Vols. 101–150 see Vol. 150

The volume numbers are printed in italics

Adam, W. and Hadjiarapoglou, L.: Dioxiranes: Oxidation Chemistry Made Easy. *164*, 45-62 (1993).
Alberto, R.: High- and Low-Valency Organometallic Compounds of Technetium and Rhenium. *176*, 149-188 (1996).
Albini, A., Fasani, E. and Mella M.: PET-Reactions of Aromatic Compounds. *168*, 143-173 (1993).
Allan, N.L. and Cooper, D.: Momentum-Space Electron Densities and Quantum Molecular Similarity. *173*, 85-111 (1995).
Allamandola, L.J.: Benzenoid Hydrocarbons in Space: The Evidence and Implications. *153*, 1-26 (1990).
Alonso, J. A., Balbás, L. C.: Density Functional Theory of Clusters of Naontransition Metals Using Simple Models. *182*, 119-171 (1996).
Anwander, R.: Lanthanide Amides. *179*, 33-112 (1996).
Anwander, R.: Routes to Monomeric Lanthanide Alkoxides. *179*, 149-246 (1996).
Anwander, R., Herrmann, W.A.: Features of Organolanthanide Complexes. *179, 1-32 (1996).*
Artymiuk, P. J., Poirette, A. R., Rice, D. W., and Willett, P.: The Use of Graph Theoretical Methods for the Comparison of the Structures of Biological Macromolecules. *174,* 73-104 (1995).
Astruc, D.: The Use of p-Organoiron Sandwiches in Aromatic Chemistry. *160*, 47-96 (1991).

Baerends, E.J., see van Leeuwen, R.: *180,* 107-168 (1996).
Balbás, L. C., see Alonso, J. A.: *182*, 119-171 (1996).
Baldas, J.: The Chemistry of Technetium Nitrido Complexes. *176*, 37-76 (1996).
Balzani, V., Barigelletti, F., De Cola, L.: Metal Complexes as Light Absorption and Light Emission Sensitizers. *158*, 31-71 (1990).
Baker, B.J. and Kerr, R.G.: Biosynthesis of Marine Sterols. *167*, 1-32 (1993).
Barigelletti, F., see Balzani, V.: *158*, 31-71 (1990).
Bassi, R., see Jennings, R. C.: *177*, 147-182 (1996).
Baumgarten, M., and Müllen, K.: Radical Ions: Where Organic Chemistry Meets Materials Sciences. *169*, 1-104 (1994).
Berces, A., Ziegler, T.: Application of Density Functional Theory to the Calculation of Force Fields and Vibrational Frequencies of Transition Metal Complexes. *182*, 41-85 (1996).
Bersier, J., see Bersier, P.M.: *170*, 113-228 (1994).
Bersier, P. M., Carlsson, L., and Bersier, J.: Electrochemistry for a Better Environment. *170*, 113-228 (1994).

Besalú, E., Carbó, R., Mestres, J. and Solà, M.: Foundations and Recent Developments on Molecular Quantum Similarity. *173*, 31-62 (1995).

Bignozzi, C.A., see Scandola, F.: *158*, 73-149 (1990).

Billing, R., Rehorek, D., Hennig, H.: Photoinduced Electron Transfer in Ion Pairs. *158*, 151-199 (1990).

Bissell, R.A., de Silva, A.P., Gunaratne, H.Q.N., Lynch, P.L.M., Maguire, G.E.M., McCoy, C.P. and Sandanayake, K.R.A.S.: Fluorescent PET (Photoinduced Electron Transfer) Sensors. *168*, 223-264 (1993).

Blasse, B.: Vibrational Structure in the Luminescence Spectra of Ions in Solids. *171*, 1-26 (1994).

Bley, K., Gruber, B., Knauer, M., Stein, N. and Ugi, I.: New Elements in the Representation of the Logical Structure of Chemistry by Qualitative Mathematical Models and Corresponding Data Structures. *166*, 199-233 (1993).

Brandi, A. see Goti, A.: *178*, 1-99 (1996).

Brunvoll, J., see Chen, R.S.: *153*, 227-254 (1990).

Brunvoll, J., Cyvin, B.N., and Cyvin, S.J.: Benzenoid Chemical Isomers and Their Enumeration. *162*, 181-221 (1992).

Brunvoll, J., see Cyvin, B.N.: *162*, 65-180 (1992).

Brunvoll, J., see Cyvin, S.J.: *166*, 65-119 (1993).

Bundle, D.R.: Synthesis of Oligosaccharides Related to Bacterial O-Antigens. *154*, 1-37 (1990).

Buot, F.A.:Generalized Functional Theory of Interacting Coupled Liouvillean Quantum Fields of Condensed Matter. *181*, 173-210 (1996)

Burke, K., see Ernzerhof, M.: *180*, 1-30 (1996).

Burrell A.K., see Sessler, J.L.: *161*, 177-274 (1991).

Caffrey, M.: Structural, Mesomorphic and Time-Resolved Studies of Biological Liquid Crystals and Lipid Membranes Using Synchrotron X-Radiation. *151*, 75-109 (1989).

Canceill, J., see Collet, A.: *165*, 103-129 (1993).

Carbó, R., see Besalú, E.: *173*, 31-62 (1995).

Carlson, R., and Nordhal, A.: Exploring Organic Synthetic Experimental Procedures. *166*, 1-64 (1993).

Carlsson, L., see Bersier, P.M.: *170*, 113-228 (1994).

Ceulemans, A.: The Doublet States in Chromium (III) Complexes. A Shell-Theoretic View. *171*, 27-68 (1994).

Clark, T.: Ab Initio Calculations on Electron-Transfer Catalysis by Metal Ions. *177*, 1-24 (1996).

Cimino, G. and Sodano, G.: Biosynthesis of Secondary Metabolites in Marine Molluscs. *167*, 77-116 (1993).

Chambron, J.-C., Dietrich-Buchecker, Ch., and Sauvage, J.-P.: From Classical Chirality to Topologically Chiral Catenands and Knots. *165*, 131-162 (1993).

Chang, C.W.J., and Scheuer, P.J.: Marine Isocyano Compounds. *167*, 33-76 (1993).

Chen, R.S., Cyvin, S.J., Cyvin, B.N., Brunvoll, J., and Klein, D.J.: Methods of Enumerating Kekulé Structures. Exemplified by Applifled by Applications of Rectangle-Shaped Benzenoids. *153*, 227-254 (1990).

Chen, R.S., see Zhang, F.J.: *153*, 181-194 (1990).

Chiorboli, C., see Scandola, F.: *158*, 73-149 (1990).

Ciolowski, J.: Scaling Properties of Topological Invariants. *153*, 85-100 (1990).

Cohen, M.H.: Strenghtening the Foundations of Chemical Reactivity Theory. *183*, 143-173 (1996).

Collet, A., Dutasta, J.-P., Lozach, B., and Canceill, J.: Cyclotriveratrylenes and Cryptophanes: Their Synthesis and Applications to Host-Guest Chemistry and to the Design of New Materials. *165*, 103-129 (1993).
Colombo, M. G., Hauser, A., and Güdel, H. U.: Competition Between Ligand Centered and Charge Transfer Lowest Excited States in bis Cyclometalated Rh^{3+} and Ir^{3+} Complexes. *171*, 143-172 (1994).
Cooper, D.L., Gerratt, J., and Raimondi, M.: The Spin-Coupled Valence Bond Description of Benzenoid Aromatic Molecules. *153*, 41-56 (1990).
Cooper, D.L., see Allan, N.L.: *173*, 85-111 (1995).
Cordero, F. M. see Goti, A.: *178*, 1-99 (1996).
Cyvin, B.N., see Chen, R.S.: *153*, 227-254 (1990).
Cyvin, S.J., see Chen, R.S.: *153*, 227-254 (1990).
Cyvin, B.N., Brunvoll, J. and Cyvin, S.J.: Enumeration of Benzenoid Systems and Other Polyhexes. *162*, 65-180 (1992).
Cyvin, S.J., see Cyvin, B.N.: *162*, 65-180 (1992).
Cyvin, B.N., see Cyvin, S.J.: *166*, 65-119 (1993).
Cyvin, S.J., Cyvin, B.N., and Brunvoll, J.: Enumeration of Benzenoid Chemical Isomers with a Study of Constant-Isomer Series. *166*, 65-119 (1993).

Dartyge, E., see Fontaine, A.: *151*, 179-203 (1989).
De Cola, L., see Balzani, V.: *158*, 31-71 (1990).
Dear, K.: Cleaning-up Oxidations with Hydrogen Peroxide. *164*, (1993).
de Mendoza, J., see Seel, C.: *175*, 101-132 (1995).
de Silva, A.P., see Bissell, R.A.: *168,* 223-264 (1993).
Descotes, G.: Synthetic Saccharide Photochemistry. *154*, 39-76 (1990).
Dias, J.R.: A Periodic Table for Benzenoid Hydrocarbons. *153*, 123-144 (1990).
Dietrich-Buchecker, Ch., see Chambron, J.-C.: *165*, 131-162 (1993).
Dobson, J. F.: Density Functional Theory of Time-Dependent Phenomena. *181*, 81-172 (1996)
Dohm, J., Vögtle, F.: Synthesis of (Strained) Macrocycles by Sulfone Pyrolysis. *161*, 69-106 (1991).
Dreizler, R. M.: Relativistic Density Functional Theory. *181*, 1-80 (1996)
Dutasta, J.-P., see Collet, A.: *165*, 103-129 (1993).

Eaton, D.F.: Electron Transfer Processes in Imaging. *156*, 199-226 (1990).
Edelmann, F.T.: Rare Earth Complexes with Heteroallylic Ligands. *179*, 113-148 (1996).
Edelmann, F.T.: Lanthanide Metallocenes in Homogeneous Catalysis.*179*, 247-276 (1996).
El-Basil, S.: Caterpillar (Gutman) Trees in Chemical Graph Theory. *153*, 273-290 (1990).
Engel, E.: Relativistic Density Functional Theory.*181*, 1-80 (1996)
Ernzerhof, M., Perdew, J. P., Burke, K.: Density Functionals: Where Do They Come From, Why Do They Work? *190,* 1-30 (1996).

Fasani, A., see Albini, A.: *168*, 143-173 (1993).
Fessner, W.-D., and Walter, C.: Enzymatic C–C Bond Formation in Asymmetric Synthesis. *184*, 97-194 (1997).
Fontaine, A., Dartyge, E., Itie, J.P., Juchs, A., Polian, A., Tolentino, H., and Tourillon, G.: Time-Resolved X-Ray Absorption Spectroscopy Using an Energy Dispensive Optics: Strengths and Limitations. *151*, 179-203 (1989).
Foote, C.S.: Photophysical and Photochemical Properties of Fullerenes. *169*, 347-364 (1994).
Fossey, J., Sorba, J., and Lefort, D.: Peracide and Free Radicals: A Theoretical and Experimental Approach. *164*, 99-113 (1993).

Fox, M.A.: Photoinduced Electron Transfer in Arranged Media. *159*, 67-102 (1991).
Freeman, P.K., and Hatlevig, S.A.: The Photochemistry of Polyhalocompounds, Dehalogenation by Photoinduced Electron Transfer, New Methods of Toxic Waste Disposal. *168*, 47-91 (1993).
Fuchigami, T.: Electrochemical Reactions of Fluoro Organic Compounds. *170*, 1-38 (1994).
Fuller, W., see Grenall, R.: *151*, 31-59 (1989).

Galán, A., see Seel, C.: *175*, 101-132 (1995).
Gehrke, R.: Research on Synthetic Polymers by Means of Experimental Techniques Employing Synchrotron Radiation. *151*, 111-159 (1989).
Geldart, D. J. W.: Nonlocal Energy Functionals: Gradient Expansions and Beyond. *190*, 31-56 (1996).
Gerratt, J., see Cooper, D.L.: *153*, 41-56 (1990).
Gerwick, W.H., Nagle, D.G., and Proteau, P.J.: Oxylipins from Marine Invertebrates. *167*, 117-180 (1993).
Gigg, J., and Gigg, R.: Synthesis of Glycolipids. *154*, 77-139 (1990).
Gislason, E.A., see Guyon, P.-M.: *151*, 161-178 (1989).
Goti, A., Cordero, F. M., and Brandi, A.: Cycloadditions Onto Methylene- and Alkylidenecyclopropane Derivatives. *178*, 1-99 (1996).
Greenall, R., Fuller, W.: High Angle Fibre Diffraction Studies on Conformational Transitions DNA Using Synchrotron Radiation. *151*, 31-59 (1989).
Gritsenko, O. V., see van Leeuwen, R.: *180*, 107-168 (1996).
Gross, E. K. U.: Density Functional Theory of Time-Dependent Phenomena. *181*, 81-172 (1996)
Gruber, B., see Bley, K.: *166*, 199-233 (1993).
Güdel, H. U., see Colombo, M. G.: *171*, 143-172 (1994).
Gunaratne, H.Q.N., see Bissell, R.A.: *168*, 223-264 (1993).
Guo, X.F., see Zhang, F.J.: *153*, 181-194 (1990).
Gust, D., and Moore, T.A.: Photosynthetic Model Systems. *159*, 103-152 (1991).
Gutman, I.: Topological Properties of Benzenoid Systems. *162*, 1-28 (1992).
Gutman, I.: Total π-Electron Energy of Benzenoid Hydrocarbons. *162*, 29-64 (1992).
Guyon, P.-M., Gislason, E.A.: Use of Synchrotron Radiation to Study-Selected Ion-Molecule Reactions. *151*, 161-178 (1989).

Hashimoto, K., and Yoshihara, K.: Rhenium Complexes Labeled with $^{186/188}$Re for Nuclear Medicine. *176*, 275-292 (1996).
Hadjiarapoglou, L., see Adam, W.: *164*, 45-62 (1993).
Hart, H., see Vinod, T. K.: *172*, 119-178 (1994).
Harbottle, G.: Neutron Acitvation Analysis in Archaecological Chemistry. *157*, 57-92 (1990).
Hatlevig, S.A., see Freeman, P.K.: *168*, 47-91 (1993).
Hauser, A., see Colombo, M. G.: *171*, 143-172 (1994).
Hayashida, O., see Murakami, Y.: *175*, 133-156 (1995).
He, W.C., and He, W.J.: Peak-Valley Path Method on Benzenoid and Coronoid Systems. *153*, 195-210 (1990).
He, W.J., see He, W.C.: *153*, 195-210 (1990).
Heaney, H.: Novel Organic Peroxygen Reagents for Use in Organic Synthesis. *164*, 1-19 (1993).
Heidbreder, A., see Hintz, S.: *177*, 77-124 (1996).
Heinze, J.: Electronically Conducting Polymers. *152*, 1-19 (1989).

Helliwell, J., see Moffat, J.K.: *151*, 61-74 (1989).
Hennig, H., see Billing, R.: *158*, 151-199 (1990).
Herrmann, W.A., see Anwander, R.: *179*, 1-32 (1996).
Hesse, M., see Meng, Q.: *161*, 107-176 (1991).
Hiberty, P.C.: The Distortive Tendencies of Delocalized π Electronic Systems. Benzene, Cyclobutadiene and Related Heteroannulenes. *153*, 27-40 (1990).
Hintz, S., Heidbreder, A., and Mattay, J.: Radical Ion Cyclizations. *177*, 77-124 (1996).
Hirao, T.: Selective Transformations of Small Ring Compounds in Redox Reactions. *178*, 99-148 (1996).
Hladka, E., Koca, J., Kratochvil, M., Kvasnicka, V., Matyska, L., Pospichal, J., and Potucek, V.: The Synthon Model and the Program PEGAS for Computer Assisted Organic Synthesis. *166*, 121-197 (1993).
Ho, T.L.: Trough-Bond Modulation of Reaction Centers by Remote Substituents. *155*, 81-158 (1990).
Holas, A., March, N. H.: Exchange and Correlation in Density Functional Theory of Atoms and Molecules. *180,* 57-106 (1996).
Höft, E.: Enantioselective Epoxidation with Peroxidic Oxygen. *164*, 63-77 (1993).
Hoggard, P. E.: Sharp-Line Electronic Spectra and Metal-Ligand Geometry. *171*, 113-142 (1994).
Holmes, K.C.: Synchrotron Radiation as a source for X-Ray Diffraction-The Beginning.*151*, 1-7 (1989).
Hopf, H., see Kostikov, R.R.: *155,* 41-80 (1990).
Houk, K. N., see Wiest, O.: *183,* 1-24 (1996).

Indelli, M.T., see Scandola, F.: *158*, 73-149 (1990).
Inokuma, S., Sakai, S., and Nishimura, J.: Synthesis and Inophoric Properties of Crownophanes. *172*, 87-118 (1994).
Itie, J.P., see Fontaine, A.: *151*, 179-203 (1989).
Ito, Y.: Chemical Reactions Induced and Probed by Positive Muons. *157*, 93-128 (1990).

Jennings, R. C., Zucchelli, G., and Bassi, R.: Antenna Structure and Energy Transfer in Higher Plant Photosystems. *177*, 147-182 (1996).
Johannsen, B., and Spiess, H.: Technetium(V) Chemistry as Relevant to Nuclear Medicine. *176*, 77-122 (1996).
John, P., and Sachs, H.: Calculating the Numbers of Perfect Matchings and of Spanning Tress, Pauling's Bond Orders, the Characteristic Polynomial, and the Eigenvectors of a Benzenoid System. *153*, 145-180 (1990).
Jones, R. O.: Structure and Spectroscopy of Small Atomic Clusters. *182*, 87-118 (1996).
Jucha, A., see Fontaine, A.: *151*, 179-203 (1989).
Jurisson, S., see Volkert, W. A.: *176*, 77-122 (1996).

Kaim, W.: Thermal and Light Induced Electron Transfer Reactions of Main Group Metal Hydrides and Organometallics. *169*, 231-252 (1994).
Kavarnos, G.J.: Fundamental Concepts of Photoinduced Electron Transfer. *156*, 21-58 (1990).
Kelly, J. M., see Kirsch-De-Mesmaeker, A.: *177*, 25-76 (1996).
Kerr, R.G., see Baker, B.J.: *167*, 1-32 (1993).
Khairutdinov, R.F., see Zamaraev, K.I.: *163*, 1-94 (1992).
Kim, J.I., Stumpe, R., and Klenze, R.: Laser-induced Photoacoustic Spectroscopy for the Speciation of Transuranic Elements in Natural Aquatic Systems. *157*, 129-180 (1990).

Kikuchi, J., see Murakami, Y.: *175*, 133-156 (1995).
Kirsch-De-Mesmaeker, A., Lecomte, J.-P., and Kelly, J. M.: Photoreactions of Metal Complexes with DNA, Especially Those Involving a Primary Photo-Electron Transfer. *177*, 25-76 (1996).
Klaffke, W., see Thiem, J.: *154*, 285-332 (1990).
Klein, D.J.: Semiempirical Valence Bond Views for Benzenoid Hydrocarbons. *153*, 57-84 (1990).
Klein, D.J., see Chen, R.S.: *153*, 227-254 (1990).
Klenze, R., see Kim, J.I.: *157*, 129-180 (1990).
Knauer, M., see Bley, K.: *166*, 199-233 (1993).
Knops, P., Sendhoff, N., Mekelburger, H.-B., Vögtle, F.: High Dilution Reactions - New Synthetic Applications. *161*, 1-36 (1991).
Koca, J., see Hladka, E.: *166*, 121-197 (1993).
Koepp, E., see Ostrowicky, A.: *161*, 37-68 (1991).
Kohnke, F.H., Mathias, J.P., and Stoddart, J.F.: Substrate-Directed Synthesis: The Rapid Assembly of Novel Macropolycyclic Structures *via* Stereoregular Diels-Alder Oligomerizations. *165*, 1-69 (1993).
Korchowiec, J. see Nalewajski, R.F.: *183*, 25-142 (1996).
Kostikov, R.R., Molchanov, A.P., and Hopf, H.: Gem-Dihalocyclopropanos in Organic Synthesis. *155*, 41-80 (1990).
Kratochvil, M., see Hladka, E.: *166*, 121-197 (1993).
Kryutchkov, S. V.: Chemistry of Technetium Cluster Compounds. *176*, 189-252 (1996).
Kumar, A., see Mishra, P. C.: *174*, 27-44 (1995).
Krogh, E., and Wan, P.: Photoinduced Electron Transfer of Carbanions and Carbacations. *156*, 93-116 (1990).
Kunkeley, H., see Vogler, A.: *158*, 1-30 (1990).
Kuwajima, I., and Nakamura, E.: Metal Homoenolates from Siloxycyclopropanes. *155*, 1-39 (1990).
Kvasnicka, V., see Hladka, E.: *166*, 121-197 (1993).

Lange, F., see Mandelkow, E.: *151*, 9-29 (1989).
Lecomte, J.-P., see Kirsch-De-Mesmaeker, A.: *177*, 25-76 (1996).
van Leeuwen, R., Gritsenko, O. V. Baerends, E. J.: Analysis and Modelling of Atomic and Molecular Kohn-Sham Potentials. *180*, 107-168 (1996).
Lefort, D., see Fossey, J.: *164*, 99-113 (1993).
Little, R. D., and Schwaebe, M. K.: Reductive Cyclizations at the Cathode. *185*, 1-48 (1997).
Lopez, L.: Photoinduced Electron Transfer Oxygenations. *156*, 117-166 (1990).
López-Boada, R., see Ludena, E. V.: *180, 169-224 (1996).*
Lozach, B., see Collet, A.: *165*, 103-129 (1993).
Ludena, E. V., López-Boada: Local-Scaling Transformation Version of Density Functional Theory: Generation of Density Functionals. *180, 169-224 (1996).*
Lüning, U.: Concave Acids and Bases. *175*, 57-100 (1995).
Lymar, S.V., Parmon, V.N., and Zamarev, K.I.: Photoinduced Electron Transfer Across Membranes. *159*, 1-66 (1991).
Lynch, P.L.M., see Bissell, R.A.: *168*, 223-264 (1993).

Maguire, G.E.M., see Bissell, R.A.: *168*, 223-264 (1993).
Mandelkow, E., Lange, G., Mandelkow, E.-M.: Applications of Synchrotron Radiation to the Study of Biopolymers in Solution: Time-Resolved X-Ray Scattering of Microtubule Self-Assembly and Oscillations. *151*, 9-29 (1989).

Mandelkow, E.-M., see Mandelkow, E.: *151*, 9-29 (1989).
March; N. H., see Holas, A.: *180,* 57-106 (1996).
Maslak, P.: Fragmentations by Photoinduced Electron Transfer. Fundamentals and Practical Aspects. *168*, 1-46 (1993).
Mathias, J.P., see Kohnke, F.H.: *165*, 1-69 (1993).
Mattay, J., and Vondenhof, M.: Contact and Solvent-Separated Radical Ion Pairs in Organic Photochemistry. *159*, 219-255 (1991).
Mattay, J., see Hintz, S.: *177*, 77-124 (1996).
Matyska, L., see Hladka, E.: *166*, 121-197 (1993).
McCoy, C.P., see Bissell, R.A.: *168*, 223-264 (1993).
Mekelburger, H.-B., see Knops, P.: *161*, 1-36 (1991).
Mekelburger, H.-B., see Schröder, A.: *172*, 179-201 (1994).
Mella, M., see Albini, A.: *168*, 143-173 (1993).
Memming, R.: Photoinduced Charge Transfer Processes at Semiconductor Electrodes and Particles. *169*, 105-182 (1994).
Meng, Q., Hesse, M.: Ring Closure Methods in the Synthesis of Macrocyclic Natural Products. *161*, 107-176 (1991).
Merz, A.: Chemically Modified Electrodes. *152*, 49-90 (1989).
Meyer, B.: Conformational Aspects of Oligosaccharides. *154*, 141-208 (1990).
Mishra, P. C., and Kumar A.: Mapping of Molecular Electric Potentials and Fields. *174*, 27-44 (1995).
Mestres, J., see Besalú, E.: *173*, 31-62 (1995).
Mezey, P.G.: Density Domain Bonding Topology and Molecular Similarity Measures. *173*, 63-83 (1995).
Michalak, A. see Nalewajski, R.F.: *183*, 25-142 (1996).
Misumi, S.: Recognitory Coloration of Cations with Chromoacerands. *165*, 163-192 (1993).
Mizuno, K., and Otsuji, Y.: Addition and Cycloaddition Reactions via Photoinduced Electron Transfer. *169*, 301-346 (1994).
Mock, W. L.: Cucurbituril. *175*, 1-24 (1995).
Moeller, K. D.: Intramolecular Carbon – Carbon Bond Forming Reactions at the Anode. *185*, 49-86 (1997).
Moffat, J.K., Helliwell, J.: The Laue Method and its Use in Time-Resolved Crystallography. *151*, 61-74 (1989).
Molchanov, A.P., see Kostikov, R.R.: *155*, 41-80 (1990).
Moore, T.A., see Gust, D.: *159*, 103-152 (1991).
Müllen, K., see Baumgarten, M.: *169*, 1-104 (1994).
Murakami, Y., Kikuchi, J., Hayashida, O.: Molecular Recognition by Large Hydrophobic Cavities Embedded in Synthetic Bilayer Membranes. *175*, 133-156 (1995).

Nagle, D.G., see Gerwick, W.H.: *167*, 117-180 (1993).
Nalewajski, R.F., Korchowiec, J. and Michalak, A.: Reactivity Criteria in Charge Sensitivity Analysis. *183*, 25-142 (1996).
Nakamura, E., see Kuwajima, I.: *155*, 1-39 (1990).
Nédélec, J.-Y., J. Périchon, and Troupel, M.: Organic Electroreductive Coupling Reactions Using Transition Metal Complexes as Catalysts. *185*, 141-174 (1997).
Nishimura, J., see Inokuma, S.: *172*, 87-118 (1994).
Nolte, R. J. M., see Sijbesma, R. P.: *175*, 25-56 (1995).
Nordahl, A., see Carlson, R.: *166*, 1-64 (1993).

Okuda, J.: Transition Metal Complexes of Sterically Demanding Cyclopentadienyl Ligands. *160*, 97-146 (1991).
Omori, T.: Substitution Reactions of Technetium Compounds. *176*, 253-274 (1996).
Ostrowicky, A., Koepp, E., Vögtle, F.: The "Vesium Effect": Synthesis of Medio- and Macrocyclic Compounds. *161*, 37-68 (1991).
Otsuji, Y., see Mizuno, K.: *169*, 301-346 (1994).

Pálinkó, I., see Tasi, G.: *174*, 45-72 (1995).
Pandey, G.: Photoinduced Electron Transfer (PET) in Organic Synthesis. *168*, 175-221 (1993).
Parmon, V.N., see Lymar, S.V.: *159*, 1-66 (1991).
Perdew, J. P. see Ernzerhof. M.: *180*, 1-30 (1996).
Périchon, J., see Nédélec, J.-Y.: *185*, 141-174 (1997).
Petersilka, M.: Density Functional Theory of Time-Dependent Phenomena.*181*, 81-172 (1996)
Poirette, A. R., see Artymiuk, P. J.: *174*, 73-104 (1995).
Polian, A., see Fontaine, A.: *151*, 179-203 (1989).
Ponec, R.: Similarity Models in the Theory of Pericyclic Macromolecules. *174*, 1-26 (1995).
Pospichal, J., see Hladka, E.: *166*, 121-197 (1993).
Potucek, V., see Hladka, E.: *166*, 121-197 (1993).
Proteau, P.J., see Gerwick, W.H.: *167*, 117-180 (1993).

Raimondi, M., see Copper, D.L.: *153*, 41-56 (1990).
Rajagopal, A. K.: Generalized Functional Theory of Interacting Coupled Liouvillean Quantum Fields of Condensed Matter. *181*, 173-210 (1996)
Reber, C., see Wexler, D.: *171*, 173-204 (1994).
Rettig, W.: Photoinduced Charge Separation via Twisted Intramolecular Charge Transfer States. *169*, 253-300 (1994).
Rice, D. W., see Artymiuk, P. J.: *174*, 73-104 (1995).
Riekel, C.: Experimental Possibilities in Small Angle Scattering at the European Synchrotron Radiation Facility. *151*, 205-229 (1989).
Roth, H.D.: A Brief History of Photoinduced Electron Transfer and Related Reactions. *156*, 1-20 (1990).
Roth, H.D.: Structure and Reactivity of Organic Radical Cations. *163*, 131-245 (1992).
Rouvray, D.H.: Similarity in Chemistry: Past, Present and Future. *173*, 1-30 (1995).
Rüsch, M., see Warwel, S.: *164*, 79-98 (1993).

Sachs, H., see John, P.: *153*, 145-180 (1990).
Saeva, F.D.: Photoinduced Electron Transfer (PET) Bond Cleavage Reactions. *156*, 59-92 (1990).
Sahni, V.: Quantum-Mechanical Interpretation of Density Functional Theory. *182*, 1-39 (1996).
Sakai, S., see Inokuma, S.: *172*, 87-118 (1994).
Sandanayake, K.R.A.S., see Bissel, R.A.: *168*, 223-264 (1993).
Sauvage, J.-P., see Chambron, J.-C.: *165*, 131-162 (1993).
Schäfer, H.-J.: Recent Contributions of Kolbe Electrolysis to Organic Synthesis. *152*, 91-151 (1989).
Scheuer, P.J., see Chang, C.W.J.: *167*, 33-76 (1993).
Schmidtke, H.-H.: Vibrational Progressions in Electronic Spectra of Complex Compounds Indicating Stron Vibronic Coupling. *171*, 69-112 (1994).
Schmittel, M.: Umpolung of Ketones via Enol Radical Cations. *169*, 183-230 (1994).

Schröder, A., Mekelburger, H.-B., and Vögtle, F.: Belt-, Ball-, and Tube-shaped Molecules. *172*, 179-201 (1994).
Schulz, J., Vögtle, F.: Transition Metal Complexes of (Strained) Cyclophanes. *172*, 41-86 (1994).
Schwaebe, M. K., see Little, R.D.: *185*, 1-48 (1997).
Seel, C., Galán, A., de Mendoza, J.: Molecular Recognition of Organic Acids and Anions - Receptor Models for Carboxylates, Amino Acids, and Nucleotides. *175*, 101-132 (1995).
Sendhoff, N., see Knops, P.: *161*, 1-36 (1991).
Sessler, J.L., Burrell, A.K.: Expanded Porphyrins. *161*, 177-274 (1991).
Sheldon, R.: Homogeneous and Heterogeneous Catalytic Oxidations with Peroxide Reagents. *164*, 21-43 (1993).
Sheng, R.: Rapid Ways of Recognize Kekuléan Benzenoid Systems. *153*, 211-226 (1990).
Sijbesma, R. P., Nolte, R. J. M.: Molecular Clips and Cages Derived from Glycoluril. *175*, 57-100 (1995).
Sodano, G., see Cimino, G.: *167*, 77-116 (1993).
Sojka, M., see Warwel, S.: *164*, 79-98 (1993).
Solà, M., see Besalú, E.: *173*, 31-62 (1995).
Sorba, J., see Fossey, J.: *164*, 99-113 (1993).
Spiess, H., see Johannsen, B.: *176*, 77-122 (1996).
Stanek, Jr., J.: Preparation of Selectively Alkylated Saccharides as Synthetic Intermediates. *154*, 209-256 (1990).
Steckhan, E.: Electroenzymatic Synthesis. *170*, 83-112 (1994).
Steenken, S.: One Electron Redox Reactions between Radicals and Organic Molecules. An Addition/Elimination (Inner-Sphere) Path. *177*, 125-146 (1996).
Stein, N., see Bley, K.: *166*, 199-233 (1993).
Stoddart, J.F., see Kohnke, F.H.: *165*, 1-69 (1993).
Soumillion, J.-P.: Photoinduced Electron Transfer Employing Organic Anions. *168*, 93-141 (1993).
Stumpe, R., see Kim, J.I.: *157*, 129-180 (1990).
Suami, T.: Chemistry of Pseudo-sugars. *154*, 257-283 (1990).
Suppan, P.: The Marcus Inverted Region. *163*, 95-130 (1992).
Suzuki, N.: Radiometric Determination of Trace Elements. *157*, 35-56 (1990).

Tabakovic, I.: Anodic Synthesis of Heterocyclic Compounds. *185*, 87-140 (1997).
Takahashi, Y.: Identification of Structural Similarity of Organic Molecules. *174*, 105-134 (1995).
Tasi, G., and Pálinkó, I.: Using Molecular Electrostatic Potential Maps for Similarity Studies. *174*, 45-72 (1995).
Thiem, J., and Klaffke, W.: Synthesis of Deoxy Oligosaccharides. *154*, 285-332 (1990).
Timpe, H.-J.: Photoinduced Electron Transfer Polymerization. *156*, 167-198 (1990).
Tobe, Y.: Strained [n]Cyclophanes. *172*, 1-40 (1994.
Tolentino, H., see Fontaine, A.: *151*, 179-203 (1989).
Tomalia, D.A.: Genealogically Directed Synthesis: Starbust/Cascade Dendrimers and Hyperbranched Structures. *165*, (1993).
Tourillon, G., see Fontaine, A.: *151*, 179-203 (1989).
Troupel M., see Nédélec, J.-Y.: *185*, 141-174 (1997).

Ugi, I., see Bley, K.: *166*, 199-233 (1993).

Vinod, T. K., Hart, H.: Cuppedo- and Cappedophanes. *172*, 119-178 (1994).
Vögtle, F., see Dohm, J.: *161*, 69-106 (1991).
Vögtle, F., see Knops, P.: *161*, 1-36 (1991).
Vögtle, F., see Ostrowicky, A.: *161*, 37-68 (1991).
Vögtle, F., see Schulz, J.: *172*, 41-86 (1994).
Vögtle, F., see Schröder, A.: *172*, 179-201 (1994).
Vogler, A., Kunkeley, H.: Photochemistry of Transition Metal Complexes Induced by Outer-Sphere Charge Transfer Excitation. *158*, 1-30 (1990).
Volkert, W. A., and S. Jurisson: Technetium-99m Chelates as Radiopharmaceuticals. *176*, 123-148 (1996).
Vondenhof, M., see Mattay, J.: *159*, 219-255 (1991).
Voyer, N.: The Development of Peptide Nanostructures. *184*, 1-38 (1997).

Walter, C., see Fessner, W.-D.: *184*, 97-194 (1997).
Wan, P., see Krogh, E.: *156*, 93-116 (1990).
Warwel, S., Sojka, M., and Rüsch, M.: Synthesis of Dicarboxylic Acids by Transition-Metal Catalyzed Oxidative Cleavage of Terminal-Unsaturated Fatty Acids. *164*, 79-98 (1993).
Wexler, D., Zink, J. I., and Reber, C.: Spectroscopic Manifestations of Potential Surface Coupling Along Normal Coordinates in Transition Metal Complexes. *171*, 173-204 (1994).
Wiest, O. and Houk, K. N.: Density Functional Theory Calculations of Pericyclic Reaction Transition Structures. *183*, 1-24 (1996).
Willett, P., see Artymiuk, P. J.: *174*, 73-104 (1995).
Willner, I., and Willner, B.: Artificial Photosynthetic Model Systems Using Light-Induced Electron Transfer Reactions in Catalytic and Biocatalytic Assemblies. *159*, 153-218 (1991).
Woggon, W.-D.: Cytochrome P450: Significance, Reaction Mechanisms and Active Site Analogues. *184*, 39-96 (1997).

Yoshida, J.: Electrochemical Reactions of Organosilicon Compounds. *170*, 39-82 (1994).
Yoshihara, K.: Chemical Nuclear Probes Using Photon Intensity Ratios. *157*, 1-34 (1990).
Yoshihara, K.: Recent Studies on the Nuclear Chemistry of Technetium. *176*, 1-16 (1996).
Yoshihara, K.: Technetium in the Environment. *176*, 17-36 (1996).
Yoshihara, K., see Hashimoto, K.: *176*, 275-192 (1996).

Zamaraev, K.I., see Lymar, S.V.: *159*, 1-66 (1991).Zamaraev, K.I., Kairutdinov, R.F.: Photoinduced Electron Tunneling Reactions in Chemistry and Biology. *163*, 1-94 (1992).
Zander, M.: Molecular Topology and Chemical Reactivity of Polynuclear Benzenoid Hydrocarbons. *153*, 101-122 (1990).
Zhang, F.J., Guo, X.F., and Chen, R.S.: The Existence of Kekulé Structures in a Benzenoid System. *153*, 181-194 (1990).
Ziegler, T., see Berces, A.: *182*, 41-85 (1996).
Zimmermann, S.C.: Rigid Molecular Tweezers as Hosts for the Complexation of Neutral Guests. *165*, 71-102 (1993).
Zink, J. I., see Wexler, D.: *171*, 173-204 (1994).
Zucchelli, G., see Jennings, R. C.: *177*, 147-182 (1996).
Zybill, Ch.: The Coordination Chemistry of Low Valent Silicon. *160*, 1-46 (1991).

Springer and the environment

At Springer we firmly believe that an international science publisher has a special obligation to the environment, and our corporate policies consistently reflect this conviction.

We also expect our business partners – paper mills, printers, packaging manufacturers, etc. – to commit themselves to using materials and production processes that do not harm the environment. The paper in this book is made from low- or no-chlorine pulp and is acid free, in conformance with international standards for paper permanency.

Printing: Saladruck, Berlin
Binding: Buchbinderei Lüderitz & Bauer, Berlin